C0-AKJ-045

Chemistry Matters!

BIOCHEMISTRY

Volume 9

Kyle Kirkland

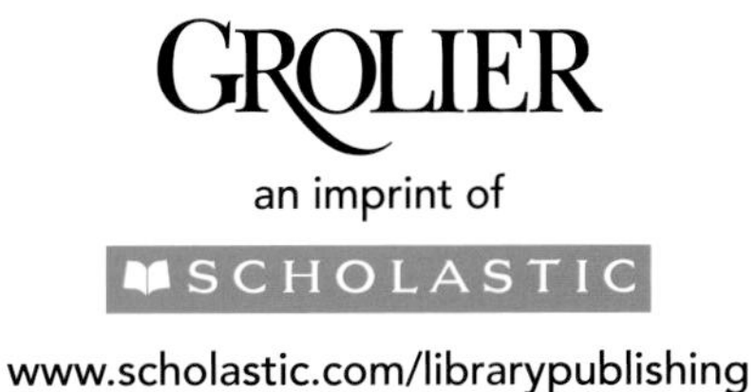

www.scholastic.com/librarypublishing

About this set

Chemistry Matters! provides an intelligent and stimulating introduction to all areas of modern chemistry as reflected in current middle school and high school curricula. This highly visual set clearly explains principles and applications using dramatic photography and annotated artwork. Carefully chosen examples make the topic fun and relevant to everyday life. Panels detail key terms, people, events, discoveries, and technologies, and include "Try This" features, in which readers are encouraged to discover principles for themselves in safe step-by-step experiments at home or school. "Chemistry in Action" boxes give everyday examples of chemical applications.

First published in 2007 by Grolier, an imprint of
Scholastic Library Publishing
Old Sherman Turnpike
Danbury, Connecticut 06816

Volume ISBN 0-7172-6203-0; 978-0-7172-6203-8
Set ISBN 0-7172-6194-8; 978-0-7172-6194-9

Library of Congress Cataloging-in-Publication Data

Chemistry matters!
v. cm.
Includes bibliographical references and index.
Contents: v.1. Atoms and molecules—v.2. States of matter—v.3. Chemical reactions—v.4. Energy and reactions—v.5. The periodic table—v.6. Metals and metalloids—v.7. Nonmetals—v.8. Organic chemistry—v.9. Biochemistry—v.10. Chemistry in action.
ISBN 0-7172-6194-8 (set : alk. paper)—ISBN 0-7172-6195-6 (v.1 : alk. paper)—ISBN 0-7172-6196-4 (v.2 : alk. paper)—ISBN 0-7172-6197-2 (v.3 : alk. paper)—ISBN 0-7172-6198-0 (v.4 : alk. paper)—ISBN 0-7172-6199-9 (v.5 : alk. paper)—ISBN 0-7172-6200-6 (v.6 : alk. paper)—ISBN 0-7172-6201-4 (v.7 : alk. paper)—ISBN 0-7172-6202-2 (v.8 : alk. paper)—ISBN 0-7172-6203-0 (v.9 : alk. paper)—ISBN 0-7172-6204-9 (v.10 : alk. paper)
1. Chemistry—Encyclopedias.
QD4.C485 2007
540—dc22

2006026209

For The Brown Reference Group plc
Project Editor: Wendy Horobin
Editors: Paul Thompson, Tim Harris, Tom Jackson, Susan Watt
Designers: Graham Curd, Joan Curtis
Picture Researchers: Laila Torsun, Helen Simm
Illustrators: Darren Awuah, Mark Walker
Indexer: Ann Barrett
Design Manager: Sarah Williams
Managing Editor: Bridget Giles
Production Director: Alastair Gourlay
Editorial Director: Lindsey Lowe
Children's Publisher: Anne O'Daly

Academic Consultants:
Dr. Donald Franceschetti, Dept. of Physics, University of Memphis
Dr. Richard Petersen, Dept. of Chemistry, University of Memphis

Printed and bound in Singapore.

Picture Credits

Front cover: Shutterstock: Mark William Penny

Alamy: Blickwinkel 6; **ARS:** Scott Bauer 66; **Corbis:** David Aubrey 45b, Fabio Cardoso/Zefa 19, Eddie Cheng/EPA 63, Chris Collins 67b, Greg English/Sygma 64, David H. Wells 42; **Nobel Foundation:** 40; **Nature Picture Library:** Pete Oxford 22; **Photos.com:** TP 5, 12, 13; **Rex Features:** Gustafssor 49, Sipa Press 57; **Shutterstock:** Galina Braskaya 29, 30, Patrick Breig 33, Stuart Elfett 31, Leon Fohn 37, Kolelnikov 43, Rohit Seth 14, Rui Vale de Sousa 17t, Ewa Walicka 20, Aaron Hitney 4, WizData Inc 48; **Science Photo Library:** 50, Andrew Lambert Photography 11l, 11r, A. Barrington Brown 27, Biophoto 38, Bsip Vem 18, Dr. Jeremy Burgess 51, 62, Dr. Tim Evans 23, Eye of Science 60, Mauro Fermariello 65, Steve Gshmesissner 17b, 24, 41, Dr. Elena Kiselcva 58, Dr. Kari Lounatmaa 45t, David McCarthy 16, David Munns 52, Pasieka 59, Mark Smith 56, Andrew Syred 9, 28, University of Newcastle on Tyne Medical School/ Simon Fraser 53, Western Opthalmic Hospital 67t; **USDA:** Larry Rana 21, Bill Tarpenning 8.

Contents

1 Carbohydrates

Biochemistry is the study of the molecules and chemical reactions in living organisms. All life on Earth depends on the chemistry of one element, carbon, and the compounds that it forms.

Elements are the basic forms of matter. All elements are made of tiny particles called atoms. When atoms of different elements group together, they form structures called molecules. Living matter uses relatively few of the 92 naturally occurring elements. In fact, most living things are made of only six: carbon, hydrogen, oxygen, nitrogen, sulfur, and phosphorus. Of the rest, only 21 are necessary for biological processes, and these occur in minute quantities called trace elements.

Carbon is a unique element because it has the capacity to form an almost infinite number of compounds

Sweet-tasting foods, including fruit, contain carbohydrates such as glucose and fructose. Carbohydrates are a source of energy for both plants and animals. Plants can make their own carbohydrates through a process called photosynthesis. Animals, however, cannot do this and must therefore eat foods that contain carbohydrates.

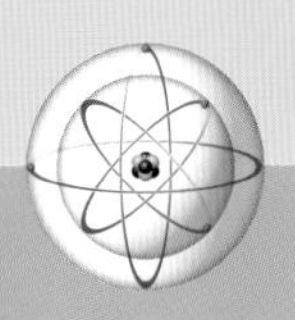

(combinations of different types of atoms). It can do this because carbon has four electrons in its outer shell. These electrons can be shared with other atoms. A whole branch of chemistry, organic chemistry, is devoted to carbon and the compounds it forms (*see* vol. 8).

Among the simplest carbon compounds used by living things are carbohydrates. Carbohydrates are the main source of energy for plants and animals. Carbohydrate molecules consist of one or more bonded saccharides (sugars). The general formula for a carbohydrate is $C_m(H_2O)_n$, where m and n are positive whole numbers greater than or equal to 3 (and can be the same number or different numbers). Carbohydrates have a backbone of carbon to which hydrogen and oxygen atoms are bonded. Because carbohydrates have a backbone of carbon atoms, they are called organic molecules. Although the general formula holds true, some complex carbohydrates have other elements such as sulfur, phosphorus, and nitrogen in their molecules.

Human cells are not able to make carbohydrates from carbon dioxide and water, so we must consume them in our diet. Plants are rich in carbohydrates because plants are able to make these molecules by a process called photosynthesis (*see* p. 43). For this reason some types of plants are excellent food sources.

Chemists divide carbohydrates into three classes. The classes are based on

Key Points

Carbohydrates get their name from their formula, $C_m(H_2O)_n$. The word "hydrate" refers to water, and carbohydrates have a number of water molecules, $(H_2O)_n$, plus a number of carbon atoms, C_m.

▼ Candy contains a lot of sucrose, a carbohydrate with a sweet taste. Sucrose is the common white sugar that you can buy as crystals for sweetening drinks such as coffee and tea.

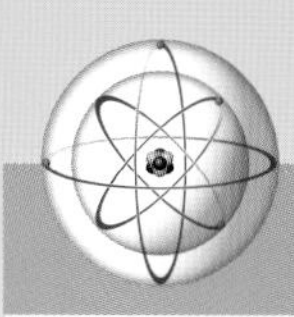

A Closer LOOK

Drawing molecules

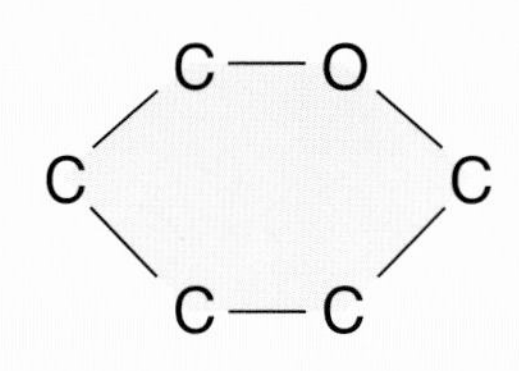

Chemists have a number of ways of drawing molecules. Here five carbon (C) atoms and a single oxygen (O) atom are connected in a ring by single bonds shown as straight lines.

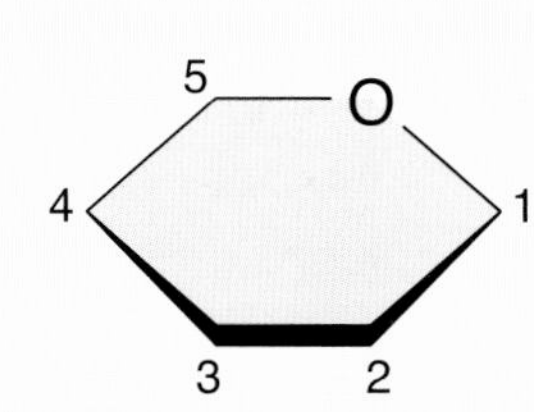

Carbon is such a common element in many molecules that sometimes the letter "C" is not shown. This diagram shows the same molecule as above. Other elements, however, such as the single oxygen atom, are still shown. The carbon atoms may also be numbered so that chemists can describe how one molecule connects to another.

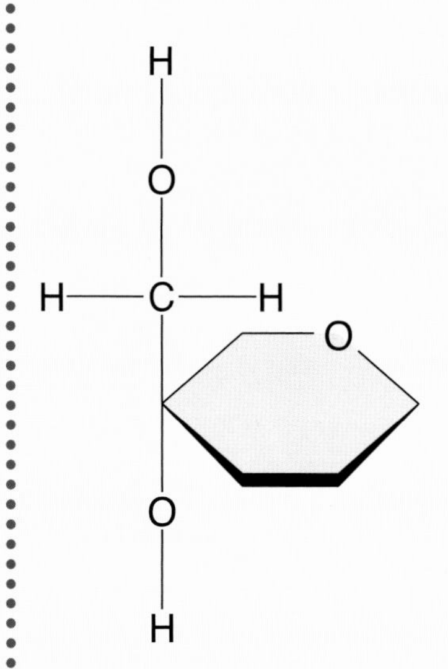

In many molecules made of rings or chains of carbon atoms, extra atoms attach to one or more of the carbons.

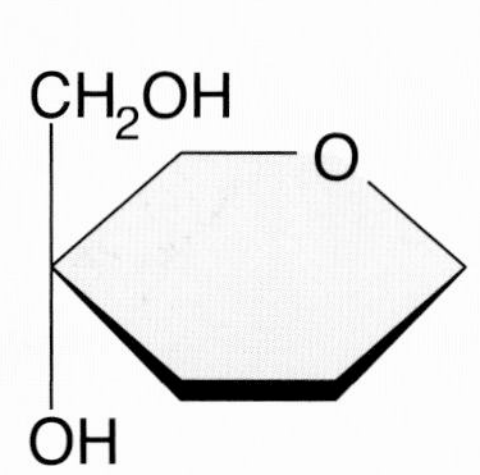

This diagram shows the same molecule as that immediately above. For simplicity the arrangement of the bonds among the additional carbon, hydrogen, and oxygen atoms is not shown. This is a common method for displaying carbohydrates.

the size of the carbohydrate molecule. Monosaccharides have a single (mono) saccharide, disaccharides have two (di), and polysaccharides have many (poly).

MONOSACCHARIDES

Monosaccharides, the simplest carbohydrates, include glucose, fructose, galactose, and others such as ribose and deoxyribose, which are components of nucleic acids (*see* pp. 25–29). The names of many sugars end in *-ose* by convention. Glucose goes by several names, including dextrose and grape sugar, and it is an

▲ *All female mammals produce milk with which they feed their young. Milk contains lactose, a disaccharide sugar made of glucose and galactose.*

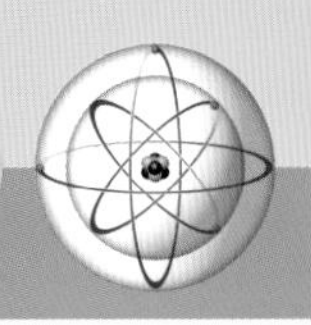

Glucose

Fructose

Galactose

▲ *In glucose, fructose, and galactose, carbon atoms form a ring with a single oxygen atom. Each of the carbon atoms is also bonded to hydrogen, carbon, and oxygen atoms in varying arrangements.*

extremely important molecule in biology. Glucose is the primary carbohydrate used as energy in humans.

Glucose, fructose, and galactose have the same formula, $C_6(H_2O)_6$, but they are different molecules. The difference is in the arrangement of their atoms. Compounds with the same chemical formula but different arrangements of their atoms are called structural isomers.

Fructose is a very sweet sugar found in table sugar, honey, and fruits. Galactose is in milk. Both fructose and galactose exist in these substances as part of a larger molecule called a disaccharide.

DISACCHARIDES

The bond joining the two saccharides of a disaccharide is called a glycosidic bond. The bond forms when an oxygen and a hydrogen (an –OH, or hydroxyl group) from each saccharide join, making a link through one of the oxygen atoms. This process releases a water molecule and so is termed a condensation reaction.

Important disaccharides include sucrose (glucose + fructose), the component of table sugar and cane sugar, and lactose (glucose + galactose), the main sugar of milk. Although milk is not sweet, it does contain sugar. Not many sugars found in nature are actually sweet. Another disaccharide is maltose, or malt sugar (glucose + glucose). In solid form, maltose is a white crystalline sugar. Maltose is sweet tasting, is found in high concentrations in germinating seeds, and is important in beer and whiskey manufacturing.

▼ *When two glucose molecules bond to form maltose, two hydrogen atoms and one oxygen atom are released, producing a molecule of water.*

Glucose

H_2O
Water

Glucose

Maltose

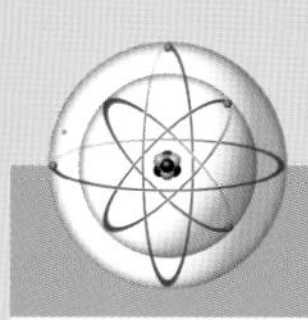

POLYSACCHARIDES

Most carbohydrates are polysaccharides, which are long chains of monosaccharides linked by glycosidic bonds. The three most common polysaccharides are starch, glycogen, and cellulose.

Carbohydrates are often ingredients of substances that cells can digest, but that is not true of the polysaccharide cellulose, at least not for human cells. Cellulose is one of the most abundant organic molecules on Earth, serving as a structural component of cell walls in plants. Cellulose forms strong, rigid fibers and these fibers give plant cells their strength and help protect them.

Cellulose consists of long, relatively unbranched (straight) chains of glucose, Typically there are a few thousand

▼ *In cellulose, chains of glucose molecules are joined by beta glycosidic bonds. These bonds occur between the first carbon of one saccharide and the fourth carbon of another saccharide.*

CH_2OH

beta glycosidic bond

glucose molecule (a saccharide)

▼ *The plants that form the bulk of a cow's diet contain a lot of cellulose. Unlike cows, some plant-eating animals are not able to digest cellulose.*

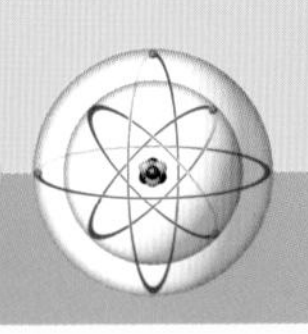

alpha glycosidic bond

In starch, alpha glycosidic bonds connect long chains of glucose molecules. Like cellulose, these bonds occur between the first and fourth carbons of adjacent saccharides. Unlike cellulose, starch is readily digested by humans.

molecules of glucose in a cellulose polysaccharide. In cellulose, glucose has what is termed a beta configuration. In starch, glucose has an alpha configuration. The difference lies in the positioning of the oxygen atom that bonds the consecutive glucose rings.

To digest cellulose, a biological molecule called an enzyme is needed to break the bonds holding the glucose molecules together. The particular type of enzyme needed is called cellulase. Humans cannot digest cellulose because we do not have cellulase in our digestive

TRY THIS

Home starch test

You can test foods for the presence of starch by this simple method.

1. Carefully grate a small portion of a vegetable such as a potato. Be careful of your fingers!

2. Put 1 tablespoon of the grated sample in ¼ cup of cold water from the faucet and stir.

3. Strain the liquid into an empty cup.

4. Add a few drops of iodine to the liquid. If the iodine turns blue–black, starch is present.

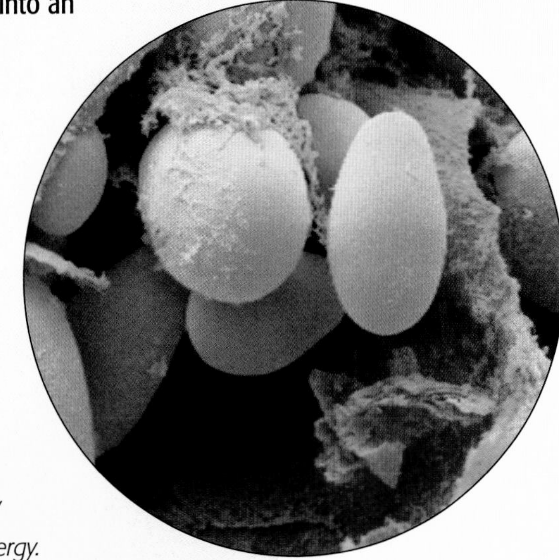

These starch grains (shaded blue) in the cell of a potato are used by the plant to store energy.

Key Terms

- **Condensation reaction:** A reaction that produces water, such as when a glycosidic bond is made.
- **Disaccharide:** Carbohydrate made of two saccharide molecules.
- **Glycosidic bond:** A bond that connects one saccharide to another saccharide.
- **Monosaccharide:** Carbohydrate made of one saccharide molecule.
- **Polysaccharide:** Carbohydrate made of many saccharide molecules.
- **Saccharide:** Sugar molecule.

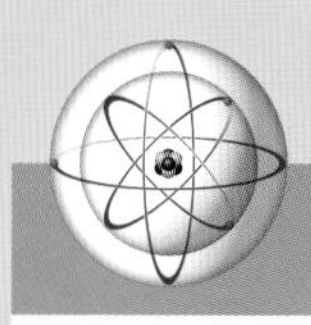

system. But termites, cows, and a few other animals harbor cellulase-containing microorganisms in their digestive tract. With the help of these microorganisms, such animals can make a meal of cellulose because their digestive systems can break it down into glucose. Even though people cannot digest cellulose and therefore can get no energy from it, the fibers are an important part of our diet because they help transport waste material through our digestive systems.

Chemistry in Action

Milk digestion

The human body cannot make use of lactose until it is broken into its saccharide components, glucose and galactose. The breaking of the glycosidic bond is catalyzed (speeded up) by a protein enzyme called lactase. Some adults lack this enzyme and are unable to break down lactose. This condition is called lactose intolerance. People who have this condition cannot consume much milk because the lactose remains in their digestive tract and can cause discomfort and diarrhea.

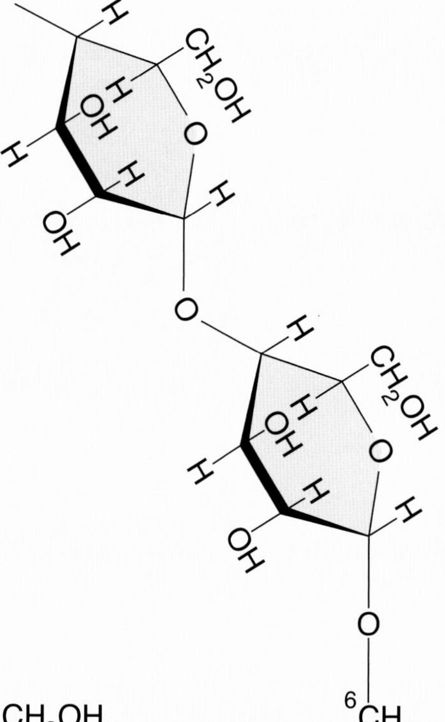

Glycogen is made of long branching chains of glucose. Here only one branch is shown. There are between 10,000 and 120,000 glucose saccharides in a glycogen molecule. The branches bond between the first and sixth carbon of the connecting saccharides.

STARCH

The primary way plants store carbohydrates for future use is as polysaccharides called starch. These polysaccharides are long chains of hundreds of glucose molecules, similar to cellulose. However, the glycosidic bonds that hold together the glucose molecules in starch have a different arrangement from the bonds in cellulose.

As far as plants are concerned, an important property of starch is that it is not soluble (it cannot dissolve) in water. Although starch takes up space in a plant cell, it does not cause the cell to draw in

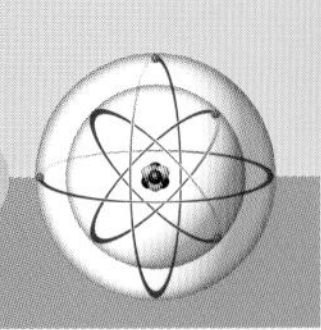

Tools and Techniques

Clockwise or counterclockwise

Some isomers (molecules that have the same chemical formula but whose atoms are arranged differently) are mirror images. The structures of these molecules, called chiral molecules, are the same except they are reversed from left to right. Many carbohydrates, such as glucose, are chiral. The two forms generally have the same properties. They can be distinguished by the arrangement of their atoms and because they rotate a type of light, called polarized light, in different directions. If an isomer rotates the light clockwise it is labeled (+), and if it rotates the light counterclockwise it is labeled (–). Rotation of polarized light is measured by an instrument called a polarimeter. In living organisms, the isomer of glucose rotates polarized light clockwise and so is labeled (+). However, the isomer of fructose in living organisms rotates polarized light counterclockwise and so is labeled (–).

The effect of passing polarized light through a solution of fructose is revealed by using a polarimeter. The color changes as the angle of the polarimeter changes.

water, which might upset the delicate water balance in the plant.

Plants store starch as an energy source for future use, but in many cases the starch ends up in human stomachs when we consume the plants. Humans can digest starch because we have the enzymes, called amylases, that break down the glycosidic bonds. Amylases work by hydrolysis, which means "breaking down with water." A hydrolysis reaction adds the water molecule lost in the condensation reaction (*see* p. 7) that formed the bond. Adding a water molecule breaks the bond.

GLYCOGEN

Plants store glucose as starch to use as needed. Animals also store glucose, but instead of making starch they make a polysaccharide from glucose called glycogen. Glycogen is stored in liver and muscle cells. Glucose is a source of energy, and the human body must maintain an adequate level of glucose in the blood to nourish cells. When the glucose level falls, which happens between meals, the liver converts glycogen into glucose. Muscle cells need a lot of energy, and glycogen provides muscles with a little extra food on demand.

See ALSO ...

- *Organic Chemistry, Vol. 8: pp. 1–65.*
- *Chemical Bonds, Vol. 3: pp. 10–21.*

2 Lipids

Lipids are greasy or waxy substances that include fats and oils. Lipids have a variety of structures and perform several different functions in living organisms. They are most important as a source of energy in the body.

It is a well-known fact that oil and water do not mix. There is a chemical reason why that is true: water is a polar molecule (has poles). Water molecules contain two hydrogen atoms and one oxygen atom joined by covalent bonds, in which the atoms share electrons. However, the oxygen and hydrogen atoms do not share electrons equally. The electrons are more attracted to the oxygen atom and pull away from the hydrogen atoms. That has the effect of making the hydrogen ends of the molecule positively charged and the oxygen end negatively charged, like the north and south poles of a magnet. Water is an excellent solvent and will dissolve many substances because its polar molecules attract other charged molecules and pull them apart (*see* vol. 2: pp. 34–45). However, lipids do not dissolve in water and are only soluble in nonpolar solvents.

NONPOLAR SOLVENTS

Nonpolar solvents are made of molecules whose atoms share electrons equally. Since the electrons are spread out uniformly there are no poles (areas

Bees secrete a wax from which they build a honeycomb. Waxes are hard lipids that do not dissolve in water.

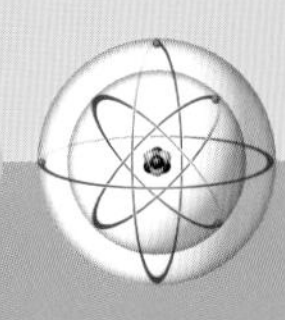

of excess charge, either positive or negative). Typical nonpolar solvents include benzene and diethyl ether.

A general rule is that nonpolar substances dissolve in nonpolar solvents. Many lipids are nonpolar, and all lipids contain some nonpolar areas in their structure. However, some lipids contain polar regions that are important to their function and behavior.

FATTY ACIDS

Fatty acids are lipids that have a chain of 12 to 24 carbon atoms with a single carboxyl group attached. A carboxyl group is made of carbon, oxygen, and hydrogen atoms and has the formula –COOH. In a saturated fatty acid, all the carbons are joined by single bonds. An unsaturated fatty acid contains one or more carbon-to-carbon double bonds. The double bond means there are fewer hydrogen atoms bonded to the carbon chain. Saturated fatty acids, which have no carbon-to-carbon double bonds, are therefore saturated (full) with hydrogen atoms. Some of the most common fatty acids have 16 carbons in their chain (for example, palmitic acid) or 18 carbons (such as oleic acid). All natural fatty acids have evenly numbered carbon chains.

Fatty acids are long chains of carbon and hydrogen atoms that have a carboxyl group at one end. Saturated fats, such as palmitic acid, have only single carbon-to-carbon bonds and form straight chains. Fatty acids that have one or more carbon-to-carbon double bonds are called unsaturated fats. They have kinks in their chain.

carboxyl group

single bond

double bond

Palmitic acid

Oleic acid

Most foods contain fatty acids. Saturated fats, such as those in the beef, cheese, and bacon of this hamburger, are bad for the body in large quantities because they are turned into substances that block blood vessels.

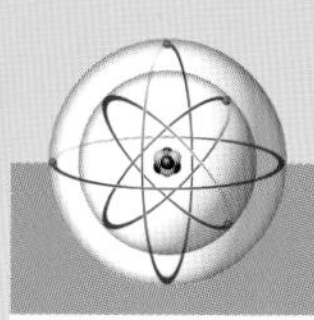

The carbon chain of saturated fatty acids is straight, so these fatty acids can stack neatly together. With their closer packing, saturated fatty acids tend to be solid at room temperature. Animal fats tend to be saturated, and therefore solid.

Double bonds between carbon molecules in unsaturated fatty acids cause bends or kinks in the chain. The bends prevent unsaturated fatty acids from packing too closely, so these fatty acids tend to be oils at room temperature. Fats from plants and many species of fish are mostly unsaturated, and therefore oily. A reaction called hydrogenation converts unsaturated fatty acids to saturated fatty acids. Hydrogenation involves blowing hydrogen through the fatty acid, which breaks the carbon double bonds and straightens the chain. Using this method liquid vegetable oil is turned into margarine, a solid.

TRY THIS

Testing for fats in food

There is a simple method to test a food for the presence of fat.

1. With a circular motion rub a small sample of the food onto a strip torn from a brown paper bag. Try cooking oil first.
2. Place a few drops of water on the bag in a location near the test sample.
3. Hold the paper up to strong light. Both water and test spot should be transparent. This is because the substances fill spaces between the paper fibers and transmit light.
4. Wait an hour or so until the water spot has dried, then hold the paper up to the light again. The water spot should no longer be transparent because the water will have evaporated. If the test spot is still relatively transparent, then it contains fat, which does not evaporate as easily as water.

TRIGLYCERIDES

Although humans and other animals store some energy as glycogen (*see* p. 9), most of the stored energy is kept in fat cells called adipocytes. The main components of fat cells are triglycerides. A triglyceride consists of a three-carbon molecule called glycerol bonded to three fatty acids. The bond, called an ester bond, forms between a hydroxyl group (–OH) of the glycerol and the –OH of the carboxyl group of a fatty acid. The three fatty acids of many triglycerides are the same, but a triglyceride may have more than one kind of fatty acid. Many of the fats and oils people eat in food are triglycerides. Steak and butter, for example, contain triglycerides.

▼ *Meat contains large molecules of fats called triglycerides. After being eaten, these fats are stored in the body in cells called adipocytes. Fats are needed to provide energy.*

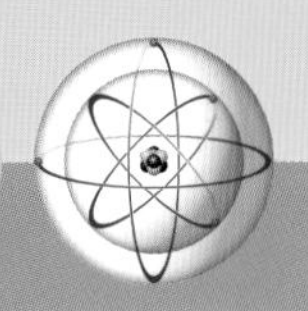

PHOSPHOLIPIDS

Triglycerides are nonpolar and do not dissolve in water. If a phosphate group ($-PO_4$), which is electrically charged, replaces one of the fatty acids, the result is a phosphoglyceride. The general name for a lipid containing a phosphate group is a phospholipid. This molecule has a nonpolar, electrically neutral region (the fatty acids) and an electrically charged polar region (the phosphate group). The oxygen atoms of the phosphate group bond with glycerol and sometimes another polar molecule.

In water, the nonpolar fatty acid "tails" of phosphoglycerides orient away from water molecules since fatty acids are hydrophobic (water hating). But the polar "head"—the phosphate group—is hydrophilic (water loving). Phospholipids in water tend to form spherical shapes as the nonpolar tails bunch together on the inside, away from the water molecules, and the polar heads form the sphere's surface. This property is useful in making structures called membranes.

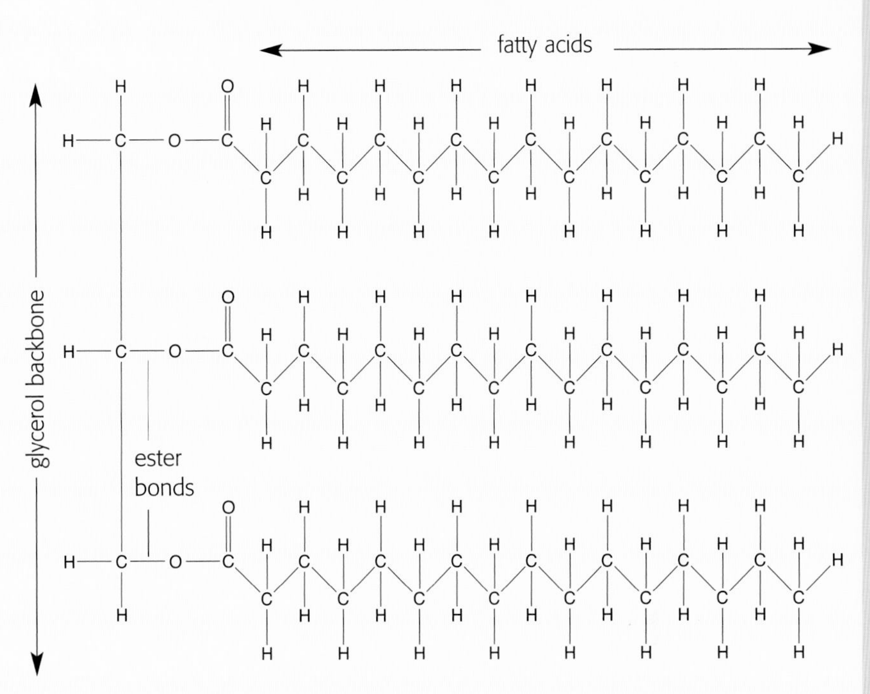

▲ *A triglyceride is made of three fatty acids connected at one end by a glycerol "backbone." In this example, triacylglycerol, the three fatty acids are all the same, but other triglycerides can have a mixture of saturated and unsaturated fatty acids.*

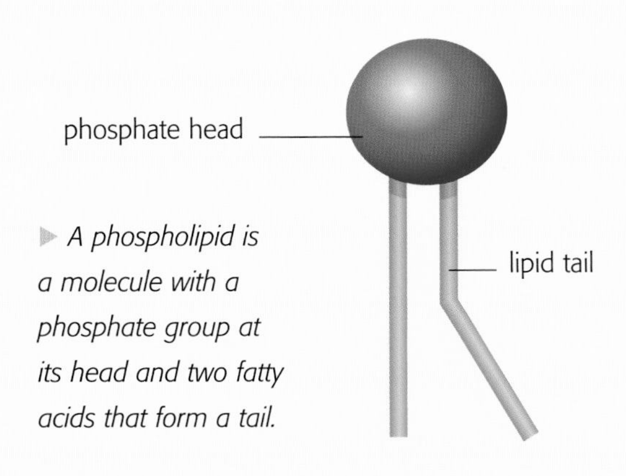

▶ *A phospholipid is a molecule with a phosphate group at its head and two fatty acids that form a tail.*

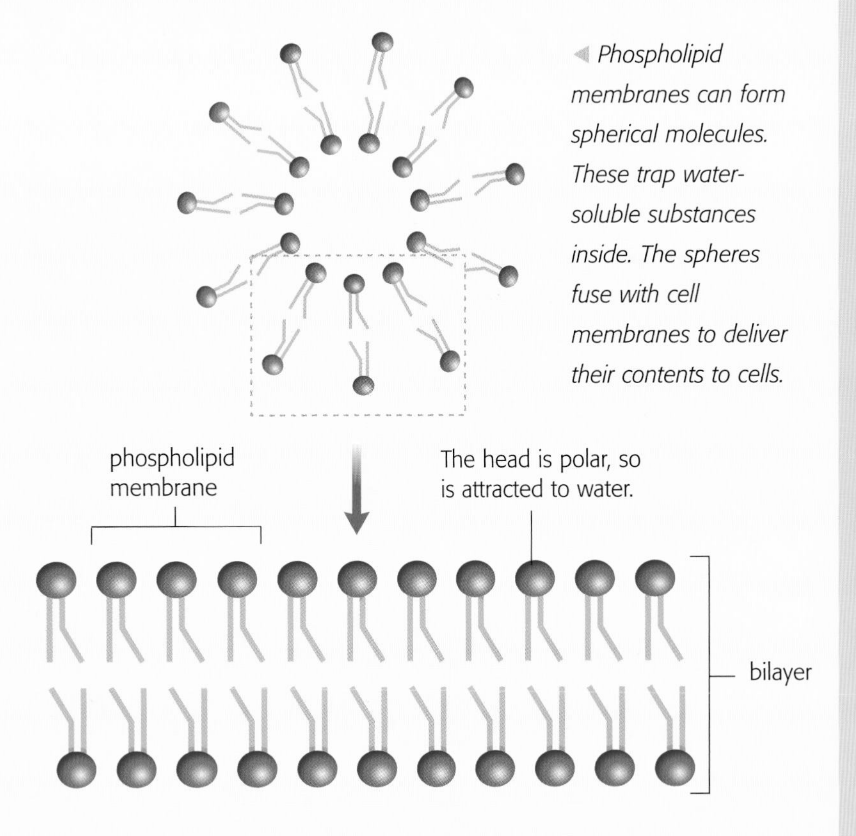

◀ *Phospholipid membranes can form spherical molecules. These trap water-soluble substances inside. The spheres fuse with cell membranes to deliver their contents to cells.*

▶ *Phospholipids can arrange themselves to form a barrier called a membrane by turning their head to face water and pointing the tail away from water.*

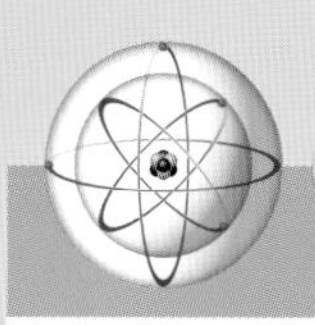

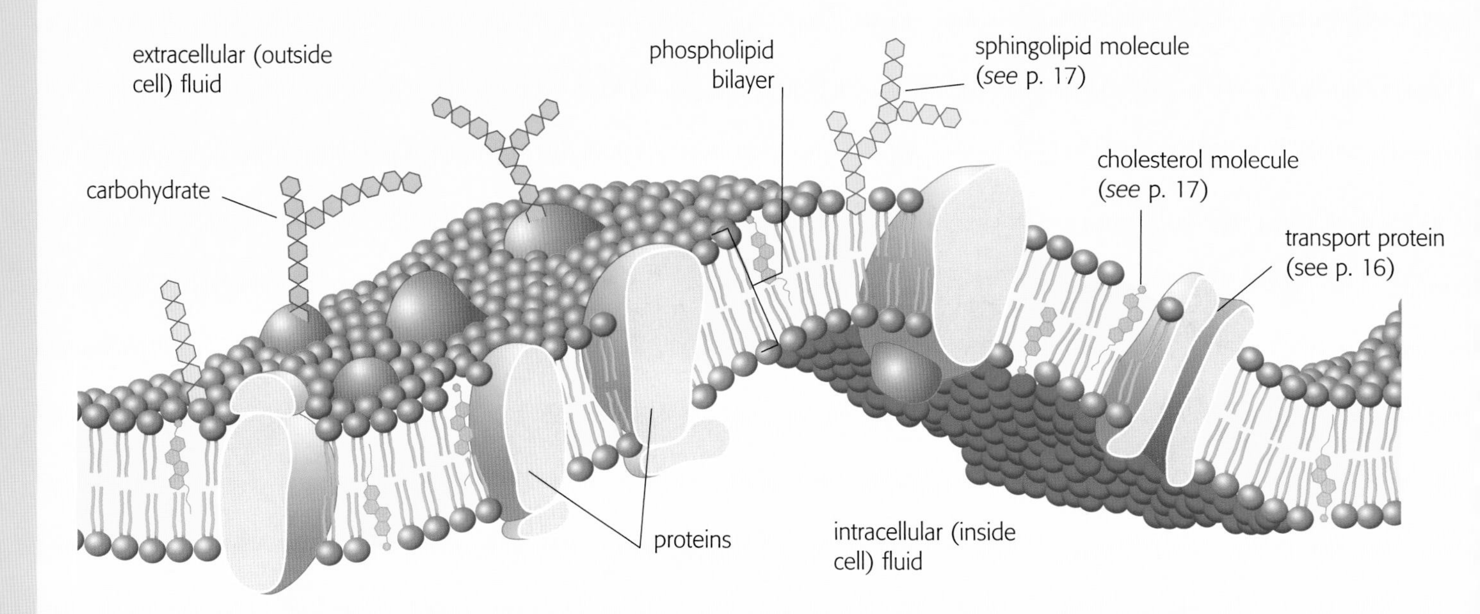

▲ *An animal cell membrane consists of a double layer of phospholipids called a bilayer. Embedded in the membrane are other molecules, such as proteins. These molecules help substances vital to the cell cross the membrane. They also keep out substances that might harm the cell.*

CELL MEMBRANES

A membrane is a thin sheet or covering. The cells of living organisms need membranes to hold in their contents, which include various nutrients and molecules as well as functional structures such as the nucleus. Membranes act as barriers, preventing the cell's contents from disappearing into the surrounding extracellular fluid. Molecules attached to the membrane or embedded inside it regulate the flow of substances into and out of the cell.

Cell membranes have two layers of phospholipids: the phospholipid bilayer. The hydrophilic (water-loving) phosphate groups form the inside and outside surfaces of the membrane and the hydrophobic (water-hating) fatty acid

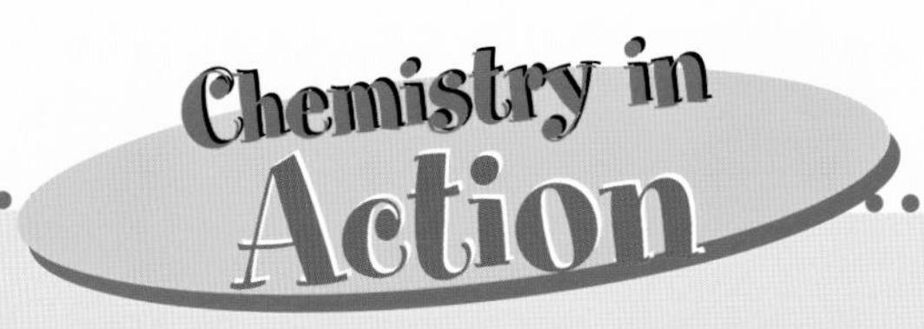

Liposomes

A liposome is a water-filled sphere made of a phospholipid bilayer, similar to a cell membrane. Physicians fill liposomes with drugs or other molecules that cannot cross cell membranes on their own. When a liposome encounters a cell membrane, the two will often join, or fuse, and the contents of the liposome spill into the cell. Liposomes thus carry medicine inside the cell.

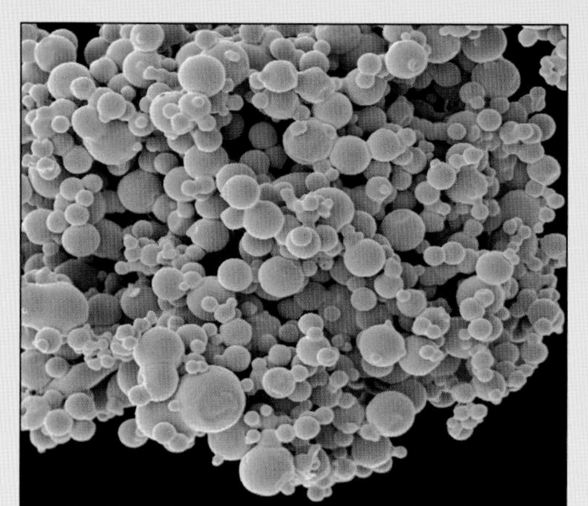

◄ *A highly magnified view of liposomes. Scientists can make liposomes in the laboratory. As well as carrying drugs into cells, liposomes are used in the cosmetic industry to fill up the wrinkles in aging skin.*

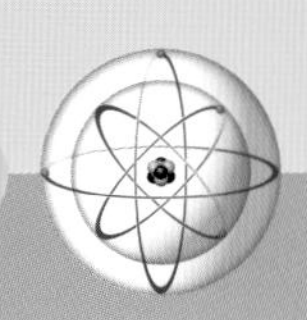

Key Terms

- **Hydrophilic:** Something that has an attraction to water.
- **Hydrophobic:** Something that does not have an attraction to water.
- **Nutrient:** A substance that provides nourishment to a cell and helps it grow or repair itself.
- **Protein:** A large biological molecule that acts as a structural component of many of the cells of living organisms.

chains form the middle of the membrane. Membranes also contain molecules such as proteins (*see* pp. 18–27) embedded within the phospholipid bilayer. The membrane is not a rigid structure because the phospholipids and embedded proteins can move or flow from one point to another. This gives the membrane a fluidlike property.

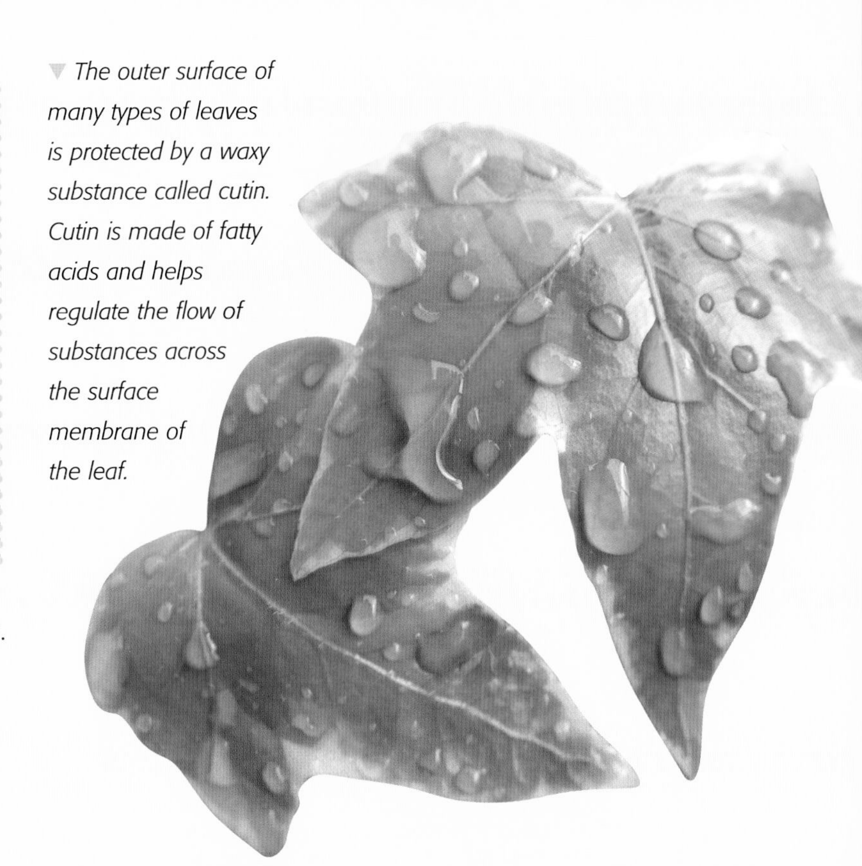

The outer surface of many types of leaves is protected by a waxy substance called cutin. Cutin is made of fatty acids and helps regulate the flow of substances across the surface membrane of the leaf.

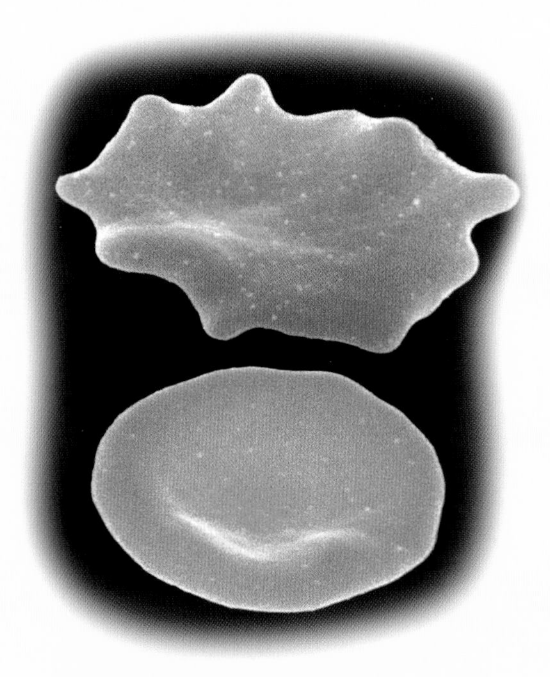

When a healthy blood cell (bottom) is exposed to a concentrated solution it loses water, shrinks, and becomes crinkly (top). This change occurs because water can diffuse across cell membranes in the process called osmosis.

MEMBRANE DIFFUSION

An important job of a cell membrane is to control and regulate the flow of molecules across it. Membranes block the passage of many substances, but water molecules are small and slip through the phospholipid bilayer by the process of diffusion. Diffusion is the ability of molecules to move about randomly until they are evenly spread out (*see* vol. 2: pp. 14–15). As water molecules roam about, they collide and move through a cell's membranes. If more water molecules move in one direction than another then there is a net flow of water. If, for example, more molecules diffuse from the outside to the inside, water inside the cell increases.

Net flows occur when concentrations are different. The fluids inside and

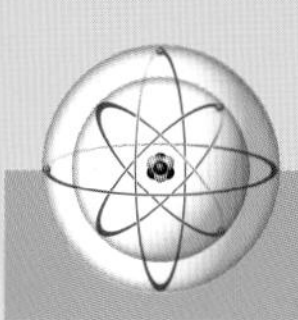

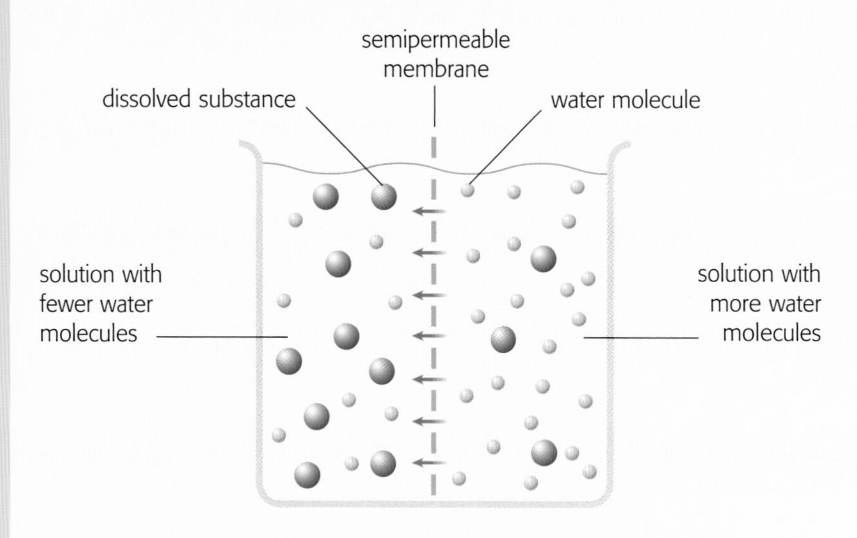

▲ *Osmosis occurs when water (yellow molecules) moves from one solution containing a dissolved substance to another. This process takes place across a semipermeable membrane. The dissolved molecules are too big to cross the membrane.*

outside of the cell contain a variety of molecules dissolved in water. Cell membranes allow only certain molecules, such as water, to permeate (cross) the membrane. Such membranes are called semipermeable. If the solution on one side of the cell membrane has a greater number of water molecules than the other, water will diffuse across the membrane to the side where there are less water molecules. Eventually the two sides reach the same concentration. The diffusion of water across a semipermeable membrane is called osmosis. Cells need to keep their interior concentrations equal to those of the exterior, otherwise there is a net flow of water across the membrane. A loss of water causes the cell to shrink, and a surplus swells it to bursting. Both situations damage the cell.

MEMBRANE TRANSPORT

Large polar molecules and certain highly charged molecules need help to cross the cell membrane. In some cases special embedded proteins called transport proteins serve as a channel so these molecules can diffuse across the membrane at certain times. Like most diffusion processes, these molecules move from areas where they are in a high concentration to areas where they are low in concentration. But sometimes the cell must transport molecules from areas of low concentration to high, against the natural flow of diffusion. This is called active transport because it needs energy to occur. This energy is derived from various metabolic processes (*see* pp. 30–45).

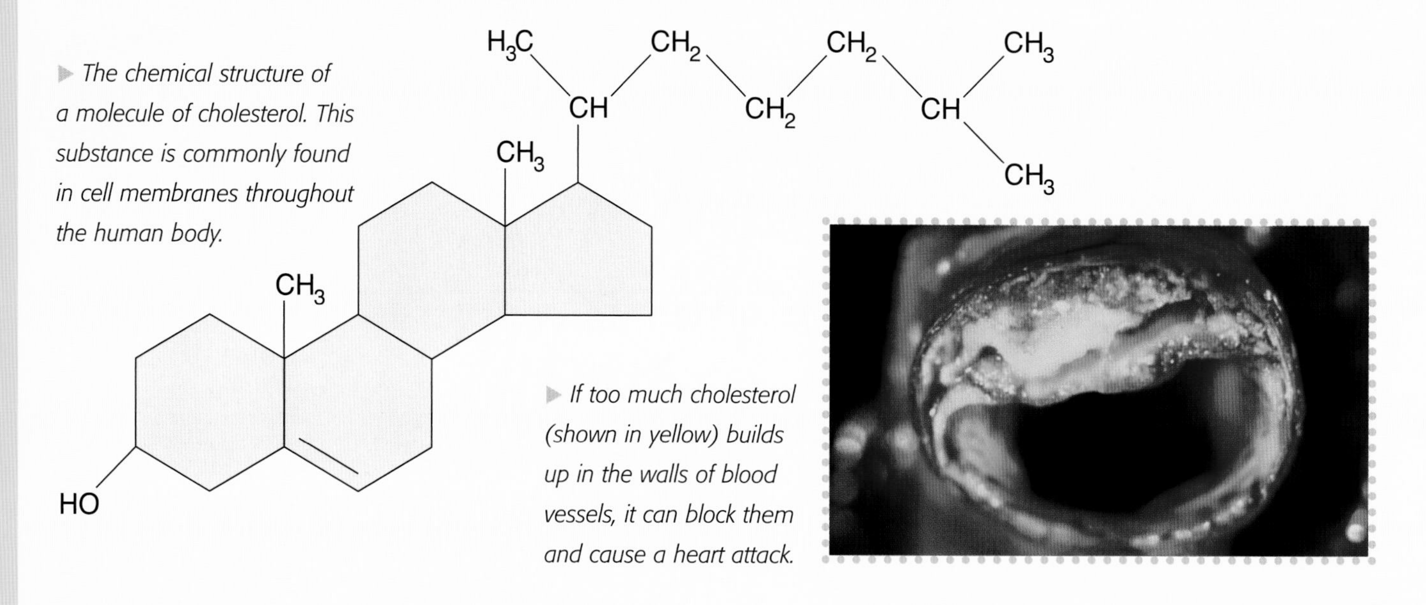

▶ *The chemical structure of a molecule of cholesterol. This substance is commonly found in cell membranes throughout the human body.*

▶ *If too much cholesterol (shown in yellow) builds up in the walls of blood vessels, it can block them and cause a heart attack.*

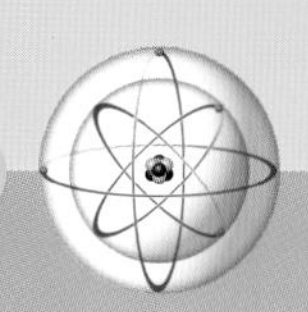

NONGLYCERIDE LIPIDS

Sphingolipids are phospholipids that do not contain glycerol. The backbone of sphingolipids is a molecule called sphingosine, a nitrogen-containing alcohol. These lipids have various functions throughout the body. One important job is to help make lipid "rafts" in the cell membranes. These rafts may serve as attachment points for membrane proteins and are important in receiving messages sent by other cells.

STEROIDS

Nonpolar molecules readily cross cell membranes because they can move in and through the nonpolar layers. Steroids are lipids that perform their function in many cases because they diffuse easily across membranes and get inside the cell. Steroids are often hormones—molecules that travel in the bloodstream and carry messages from cells in one part of the body to another. The structure of steroids is entirely different than that of triglycerides and phospholipids; steroids consist of rings of carbon fused to resemble a mesh fence.

Steroid hormones are derived from an important steroid lipid called cholesterol. Cholesterol occurs in cell membranes and can also travel through the blood, often attached to complex combinations of lipids and proteins. The liver makes cholesterol, but the body also obtains it from food. Cholesterol is an essential component to the body but excesses of this molecule have been linked to medical disorders such as heart disease and gallstones.

A Closer LOOK

The dangers of steroids

Anabolic steroids are artificial hormones that help muscle growth. Sometimes athletes inject a large amount of anabolic steroids into their bodies in an attempt to increase muscle size and improve their performance. This is a dangerous practice that can result in mental problems and serious physical injuries. Most athletic contests, such as the Olympics, ban these substances.

Athletes build muscle naturally through exercise. However, some take artificial steroids that help speed up this process. A number of athletes have been banned from competition after using steroids.

See ALSO ...
- *Other Organic Compounds, Vol. 8: pp. 45–48.*

3 Proteins and Nucleic Acids

The albumen, or egg white, of a chicken's egg contains a relatively high proportion of protein. This includes all nine "essential amino acids," those that are not produced by the human body.

Proteins and nucleic acids are the most important molecules in the body. They are used as structural molecules to build skin, hair, and muscle. They are also vital for the way the body functions and reproduces cells.

Biochemical molecules are generally large and complex, but they are usually composed of much simpler units that are bonded together. Proteins, for example, are strings of amino acids. The word *protein* means "of first importance." Some proteins called enzymes speed up chemical reactions so they occur quickly enough to support life. Other proteins, such as keratin, form rigid structures like horns or fingernails. Regardless of their function, all proteins are made of a sequence of amino acids.

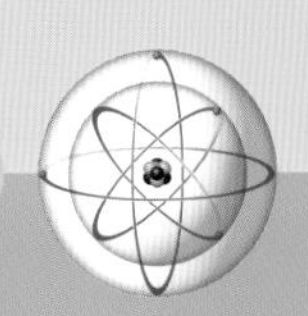

AMINO ACIDS

Amino acids are so named because they are made from two groups of atoms, an amino group and a carboxylic acid. Although there are more than 200 types of amino acids, the proteins in most living organisms are made from only 20 amino acids. These are: alanine, arginine, asparagine, aspartate (or aspartic acid), cysteine, glutamate (or glutamic acid), glutamine, glycine, histidine, isoleucine, leucine, lysine, methionine, phenylalanine, proline, serine, threonine, tryptophan, tyrosine, and valine. Different proteins can have different numbers of amino acids. Small proteins, called peptides, can have fewer than 20 amino acids. One of the

A Closer LOOK

What is an amino acid?

An amino acid consists of a central ("alpha") carbon atom with four attachments. These are:

1) a hydrogen atom
2) a carboxyl group, –COOH (in water this acidic group often loses its positively charged hydrogen atom and becomes COO^-, a carboxylate ion)
3) an amino group, $-NH_2$ (in water this group often gains a hydrogen nucleus and becomes NH_3^+)
4) a side chain, usually designated as R, R_1, R_2, and so on.

Amino acids differ from each other because they have different side chains. For example, the side chain for glycine (*see* p. 22) is just –H, and methionine is $-CH_2-CH_2-S-CH_3$.

The horns of this goat, and its fur, are made of tough proteins called keratins. There are many different kinds of keratins, but all contain a high proportion of the two smallest amino acids, glycine and alanine.

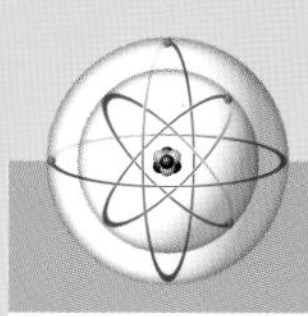

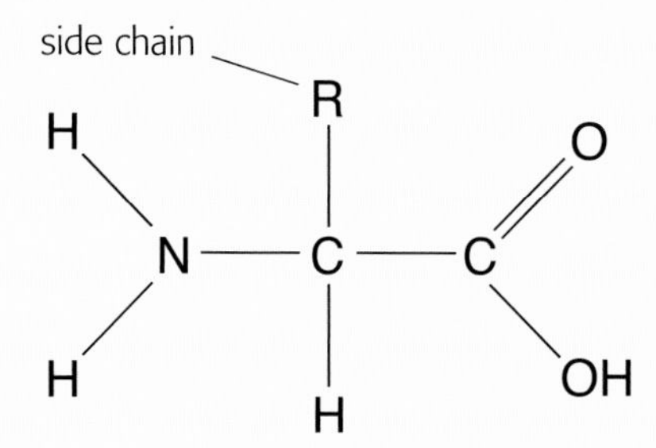

Glycine is the simplest amino acid, and its side chain is made up of a single hydrogen atom. Most proteins contain only small quantities of glycine.

largest proteins, titin, which is used in muscle contraction, is a chain of about 27,000 amino acids.

PEPTIDES

Peptides are formed from two amino acids bonded together. A covalent bond (*see* vol. 1: p. 48) joins the carboxyl group of one amino acid and the amino group of another. Proteins are made of a sequence of peptides. A peptide bond forms by a condensation reaction, which is a reaction that releases water (H_2O). The water molecule comes from an OH of the carboxyl group of one amino acid and an H of the amino group of the other amino acid. The side chains are not involved in the reaction. Amino acids contain mostly carbon (C), hydrogen (H), oxygen (O), and nitrogen (N). (Cysteine and methionine also have sulfur.) These elements are the most important in biochemistry, making up about 95 percent of the human body.

Chemistry in Action

Deadly proteins

Snake venom is a mixture of hundreds of proteins. Some of the proteins are toxins (poisons), useful for hunting and defense. Snakes use their fangs to inject venom into a victim's bloodstream. In humans, these toxins can disrupt the activity of the muscles, brain, heart, or other tissues, sometimes even causing death.

Snakes inject venom through two long fangs on the upper jaw. The venom of vipers, such as this pit viper, clots the arteries of the victim.

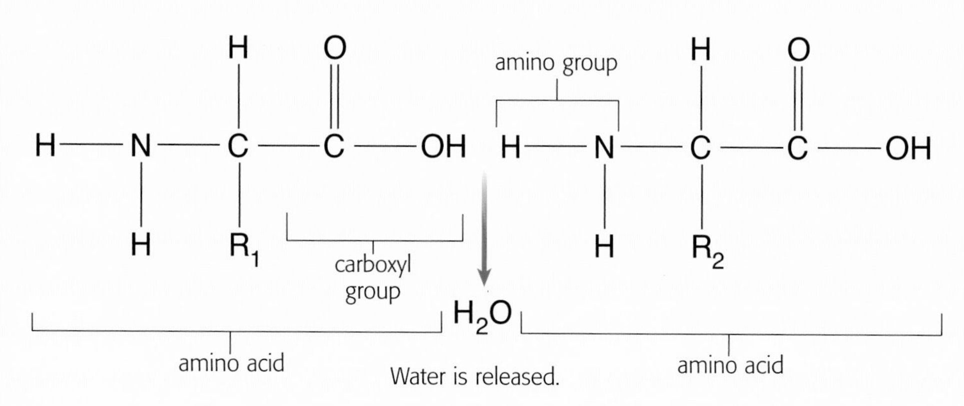

A peptide is formed when the carboxyl group of one amino acid bonds with the amino group of another. Water is released during the reaction. Proteins are made up of sequences of peptides.

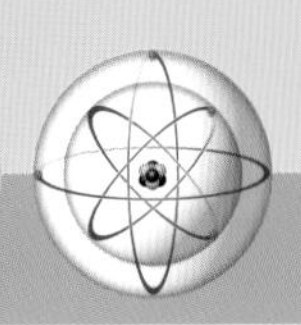

SEQUENCE

The structure of a protein is critical to its function. Some proteins are flat and some fold up into balls. Yet all proteins have a three-dimensional structure that is necessary for them to do their job. The structure derives from the protein's sequence of amino acids.

The structure of a protein has four levels—primary, secondary, tertiary, and quaternary. The sequence of amino acids is the primary structure, and this governs the shape of the protein. Interactions among the amino acids form weak hydrogen bonds. These bonds determine and maintain the protein's shape, or secondary structure. The position (sequence) of the amino acids dictates where these bonds occur. The bonds twist the protein to form a helical segment called an alpha helix or a flattened segment, or beta pleated sheet.

A protein's tertiary structure reflects the way the alpha helices and beta sheets

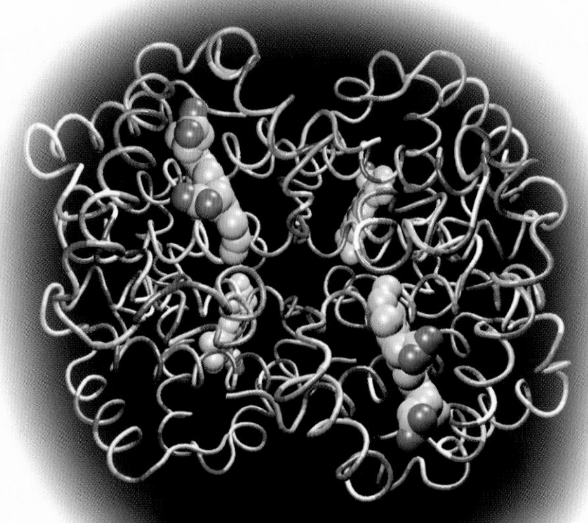

◀ *This is a computer-generated image of hemoglobin, a protein in human blood. Hemoglobin plays a crucial role in respiration. It binds with oxygen, carrying it from the lungs to the body's tissues. It also carries carbon dioxide (CO_2) from the tissues back to the lungs.*

▶ *An alpha helix is a right-handed coil with an N–C–C backbone. It is one of the most common secondary structures in proteins.*

▼ *The other common secondary structure is a beta pleated sheet. Both structures owe their shapes to hydrogen bonds between different parts of the same chain.*

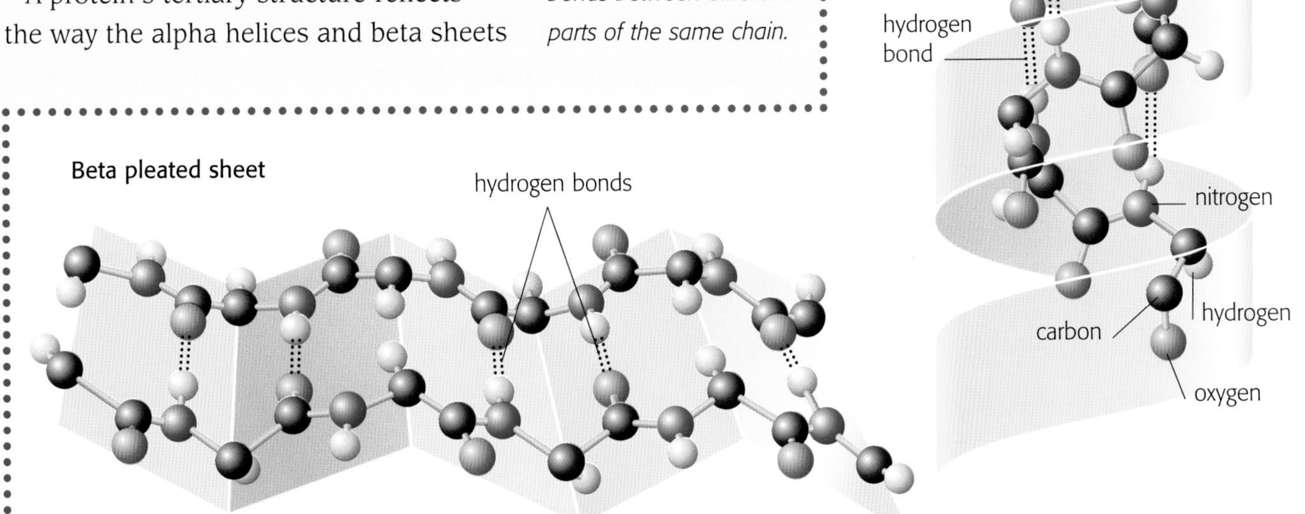

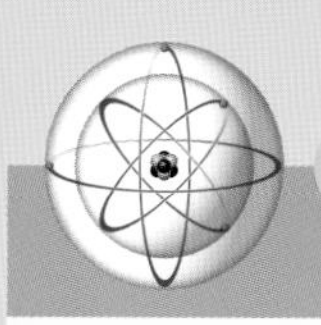

fold into the protein's shape. In some proteins this shape is modified by separate chains of amino acids, each with their own tertiary structure. These amino acids change the molecule again to give the final quaternary structure.

Although proteins are large molecules, they are much too small to be seen, even with powerful microscopes. To determine a protein's shape scientists often use X-ray crystallography. X-rays are the high-frequency radiation that doctors use to make images of bones and other body parts. Biochemists use X-rays to make images of proteins.

Tools and Techniques

X-ray crystallography

X-ray crystallography requires an X-ray source, an X-ray detector, and a crystal of the protein to be studied. A crystal contains many molecules in a regular, geometric arrangement. Some proteins crystallize easily but others do not, so the shape of some proteins is easier to determine than others. The X-rays strike the crystal and scatter, altering their path. The arrangement of atoms affects this scattering. Scientists can determine a protein's shape by studying the way the X-rays scatter.

FUNCTION

Many proteins are soluble in water. Most of these proteins are globular (globelike) in shape and perform various functions. An example is hemoglobin, a protein in the blood of mammals. Hemoglobin's job is to carry oxygen to nourish the cells of the body. Hemoglobin has four subunits (separate chains) and contains four iron atoms each capable of binding oxygen.

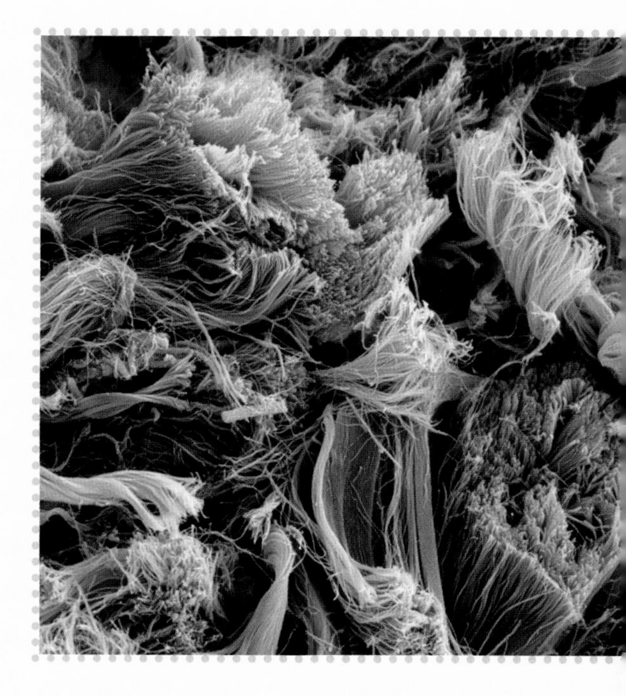

Collagen is the most common protein in the body of a mammal. Collagen is the main component of bones and teeth. It also strengthens skin and blood vessels.

Chemistry in Action

Folding proteins

A protein folds itself, sometimes with help from other molecules. However, extreme heat breaks the weak bonds holding the shape together. When these weak bonds are broken, the protein fails to function and is said to be denatured. This is one of the reasons why animals and people cannot live in very hot environments—their proteins break down and stop working.

Other proteins are insoluble in water. These proteins usually consist of long sheets or fibers. An example is collagen, a common protein; about one-third of all protein in the human body is collagen. Collagen strengthens skin and other tissues. Collagen's insoluble nature is essential for its function—our skin would be of little use if it dissolved in rain!

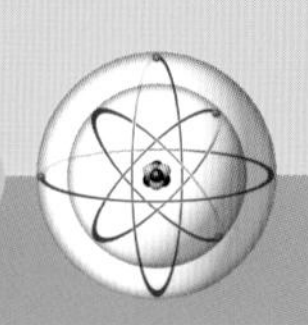

Key Terms

- **Amino acid:** A component of a protein made up of a carboxyl group (–COOH) and an amino group ($-NH_2$), both attached to the same C atom.
- **Keratin:** A type of protein that forms structures such as fingernails and horns.
- **Nucleotide:** A molecule made up of a nitrogen-containing base, a sugar, and a phosphate group.
- **Protein:** A large biological molecule made of amino acids. Proteins act as structural components of many cells.

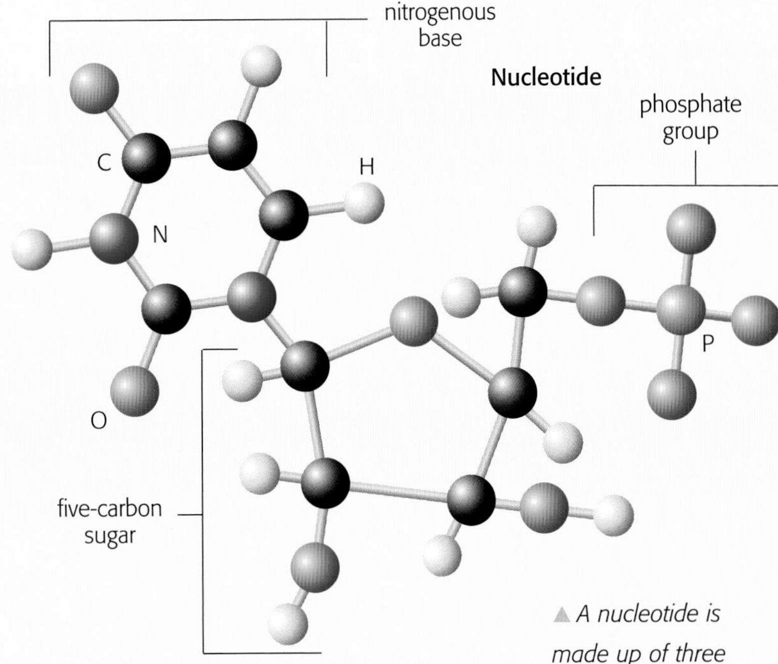

A nucleotide is made up of three main components: a base containing nitrogen; a sugar; and a phosphate group, with one phosphorus atom and four oxygen atoms.

NUCLEIC ACIDS

Carbon is vital in biochemistry because it forms long chains of molecules. Proteins are one important group of biochemical molecules. Another important group of molecules are nucleic acids.

Nucleic acids are strings of chemically bonded molecules called nucleotides. Nucleotides are made of three basic components: a phosphate group, which consists of one phosphorus (P) atom and four oxygen (O) atoms (PO_4^{3-}); a five-carbon sugar such as ribose ($C_5H_{10}O_5$); and a base containing nitrogen. This base can belong to one of two types, purines or pyrimidines. A purine has a five-atom ring structure connected to a six-atom ring, and a pyrimidine has a six-atom ring. Nucleic acids store and transfer information that every cell must have to function. There are two major types of nucleic acids: ribonucleic acid (RNA) and deoxyribonucleic acid (DNA).

Adenine

Guanine

PURINES

Nitrogenous bases called purines are components of nucleotides. Adenine and guanine are two kinds of purines.

Thymine

Uracil

Cytosine

PYRIMIDINES

Pyrimidines are also components of nucleotides.

RNA

The sugar of RNA nucleotides is ribose (which gives RNA its name) and the bases are one of the following: adenine (A) and guanine (G), both of which are purines, and cytosine (C) and uracil (U), both of which are pyrimidines. RNA molecules are usually single strands of nucleotides connected by a condensation reaction (*see* p. 22). The most important aspect of RNA structure is the sequence of bases.

Nucleic acids are named for their association with a cell's nucleus—the structure inside a cell that holds genetic (inherited) information. RNA does not always stay in the nucleus. Three classes of RNA molecules exist: messenger RNA (mRNA) carries information from the nucleus into the cell; outside the nucleus, ribosomal RNA (rRNA) helps make new proteins using mRNA; and transfer RNA (tRNA) helps translate the information carried by mRNA.

▶ *The structure of a molecule of DNA. On either side is a backbone of sugar and phosphate molecules. A base is attached to each of these. The bases can be adenine (A), guanine (G), thymine (T), or cytosine (C). Each base is linked to its partner on the other side by hydrogen bonding. A always bonds with T, and G always bonds with C.*

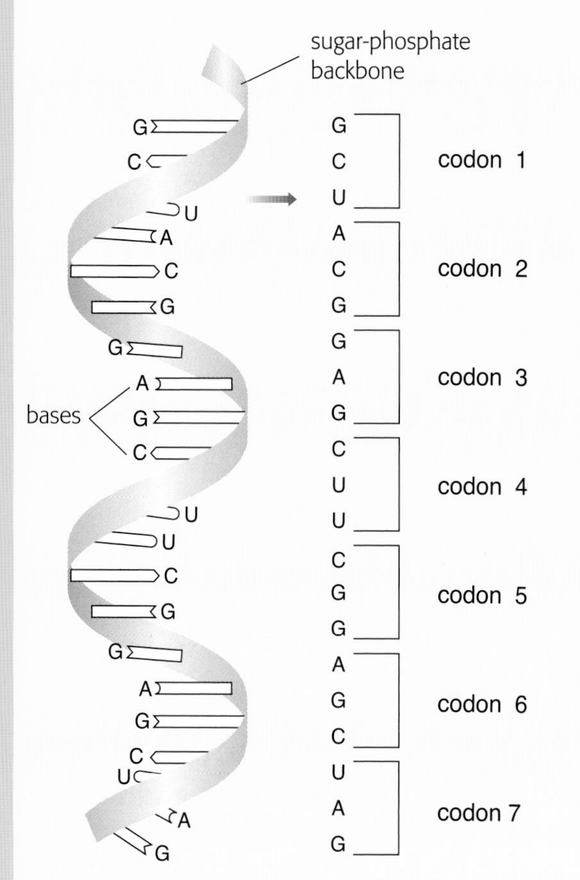

◀ *RNA carries genetic (inherited) information that controls how a cell functions and reproduces. It does this by dictating which proteins are made. RNA is made up of nucleotides composed of a phosphate and a sugar, which form the molecule's backbone, and one of four bases. The nucleotides are read in groups of three called codons. Each triplet code, such as CUU, is the code to make a different amino acid.*

DNA

The sugar of DNA nucleotides is deoxyribose, the same as ribose except that one oxygen atom is missing. The bases are the same as RNA with the exception of uracil, which is replaced in DNA by thymine (T). Long strands of nucleotides form DNA molecules.

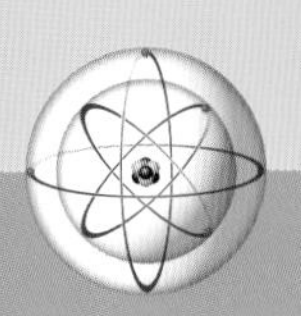

Profile

Watson and Crick

On April 25, 1953, James Watson (1928–) and Francis Crick (1916–2004) published a short paper in the journal *Nature* describing DNA's double-helix structure. Their work was based on X-ray crystallography by Maurice Wilkins (1916–2004) and Rosalind Franklin (1920–1958). Watson, Crick, and Wilkins shared the 1962 Nobel Prize for physiology. (Franklin died earlier, and Noble Prizes are awarded only to living scientists because the purpose of the prizes is to stimulate ongoing research.)

▶ *James Watson (left) and Francis Crick in 1953, with their model of the structure of the DNA double helix.*

Scientists discovered in the 1940s that DNA in the cell's nucleus holds genetic information, but no one knew how. Then James Watson and Francis Crick discovered the normal structure of DNA is a double helix. Single strands of DNA molecules twist around each like a spiraling ladder.

The double helix forms because bases on each strand form weak bonds. A pyrimidine base on one strand bonds with a purine base on another. The double helix is a stable molecule, allowing DNA to exist for long periods without breaking down.

GENES

Segments of DNA and the sequence of bases they carry form genes. Some DNA sequences are genes, and other sequences regulate access to this information. Each double-stranded DNA helix coils around itself and forms a single chromosome in the cell's nucleus. If stretched out, the DNA in a single human cell would be 6 feet (1.83 m) long. Organisms have various numbers of chromosomes—humans have 23 pairs, mice 20 pairs, and chimpanzees 24

A Closer LOOK

Zipping up a molecule

A pyrimidine base forms a bond called a hydrogen bond (*see* p. 23) with a purine base. Because of the double-helix structure and the geometry of the nucleotide bases, C always bonds with G, and A bonds with T. The bonds "zip up" the double-helix molecule. But hydrogen bonds are not covalent (electron sharing) bonds and are easily broken. The ability of a DNA molecule to "unzip" is just as important as the ability to zip up. DNA has to unzip to pass on its genetic information to molecules of RNA.

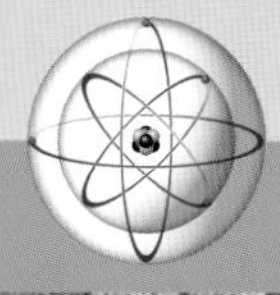

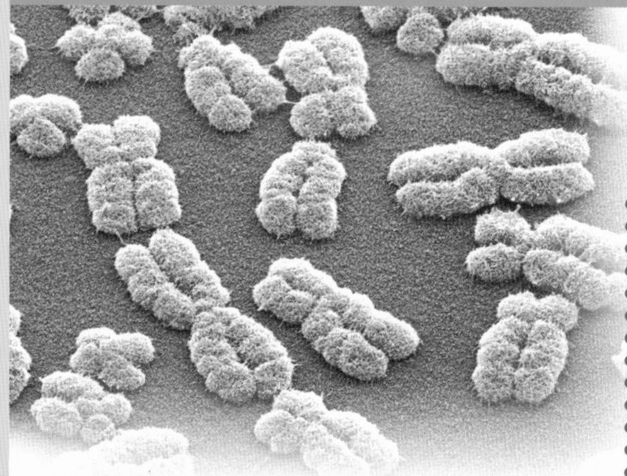

An organism's genetic information is packaged in large molecules called chromosomes. Humans have 23 pairs of chromosomes.

Chemistry in Action

Catching criminals

DNA fingerprinting is a technique to identify an individual based on his or her DNA. Crime investigators called forensics experts gather samples such as blood, skin, and sweat from a crime scene. They test the samples in an attempt to identify the criminal. If a suspect's DNA is found in the sample, it shows that he or she was at the scene of the crime, as would a fingerprint.

TRY THIS

Fruity DNA

Materials: a large bowl of ice, methylated spirits (available at hardware stores), a kiwi fruit, a knife and chopping board, measuring cups, kitchen scales, table salt, a measuring cylinder, dishwashing liquid, water, a saucepan or bowl of hot water, a strainer, and a large spoon

1. Peel and cut the kiwi into small pieces and place in measuring cup.
2. Stir 0.1 ounce (3 g) of salt, 0.33 fluid ounces (10 ml) of dishwashing liquid, and 3 fluid ounces (100 ml) of water and add to the measuring cup, letting the mixture sit for 15 minutes.
3. Put the measuring cup in a pan of hot water for 15 minutes.
4. Pour the resulting green liquid in the measuring cup through a kitchen strainer and into a glass.
5. Carefully pour ice-cold methylated spirits on the back of a spoon over the glass, forming a purple layer of the same width on top of the green layer. Leave to stand for at least 30 minutes.

   WARNING: Methylated spirits are dangerous and should never be consumed!

6. Kiwi DNA should appear as a white layer between the green and purple liquid. You can fish it out with a loop of wire or a fork.

The DNA of the kiwi fruit is visible as a thin white layer between the greenish liquid below and the purple layer above.

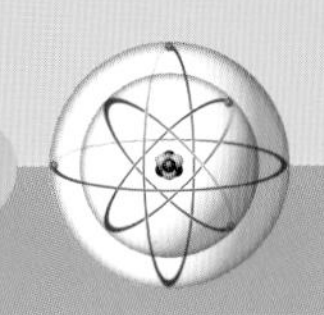

pairs. One of each chromosome pair comes from the mother and one from the father.

Every species (type) of animal and plant has a unique set of genes that defines the structure and function of their cells, tissues, and organs. But many genes come in slightly different sequences called alleles, and each individual within a species may have a different set of alleles. Genetic differences account for much of the variation in appearance and behavior of species and of individuals within species.

EVOLUTION

Different alleles give different characteristics. In some situations one allele may have advantages, such as letting an animal move faster. However, the same allele may require the animal to find more food. Animals better suited to their environment tend to survive and pass along their gene sequences, a process called natural selection. This process is the basis of evolution.

Species evolve as conditions change. The DNA of individuals varies as a result of several factors, including mutations. Mutations are accidental changes in DNA that are not related to the environment in which the organism lives. However, they may make the animal or plant better suited to its environment. If so, there is more chance that its DNA will be passed on to the next generation. Over a long time, an accumulation of mutations can lead to the evolution of a new species. Reproduction mixes genes, resulting in more genetic variability from which natural selection chooses a "winner."

A Closer LOOK

Dominant genes

Individuals have two copies (alleles) of each gene, one on each chromosome pair. In some cases each of the two is the same sequence, but in some cases the sequence of the gene is different. When the two copies differ then one or the other will often (though not always) be dominant, and the trait caused by the gene will come from this particular sequence.

HUMAN GENOME PROJECT

In 2003 scientists completed a major goal of the Human Genome Project, sequencing all human genes—the human genome. After extracting DNA from a few people, large machines automatically sequenced the 3 billion bases, abbreviated C, G, A, and T, of genetic information. Since genes influence many characteristics, including susceptibility to cancer and heart disease, the Human Genome Project will give scientists and physicians valuable insights into human biology.

See ALSO ...
- *Making Molecules, Vol. 9: pp. 46–65.*

All dogs are members of the same species, but no two are exactly the same. Differences in size and fur color reflect differences in an individual's DNA.

4 Metabolic Pathways

Metabolism refers to changes brought about by the many chemical reactions occurring in the cells of living organisms. Many of the reactions follow a series of steps called a metabolic pathway.

One of the most important changes produced by metabolism is the conversion of food to energy. People need energy to run and jump and even to do very gentle activities such as reading and thinking. Cells, of course, need energy to support these activities as well as many others. This energy comes from food, which must be broken down by the body before it can be used.

Our bodies need energy to play tennis or perform any other function. This energy is produced by chemical reactions that occur after the food we eat has been broken down, or digested. The sum of all these reactions is called the body's metabolism.

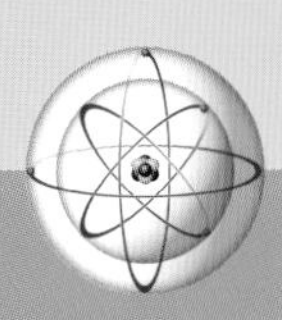

ENERGY

Energy is always conserved (cannot be lost) but it can be transformed into other forms of energy (*see* vol. 4: 4–17). Potential energy is stored energy, while kinetic and thermal (heat) energy are energies of motion. Chemical reactions that produce thermal or kinetic energy from potential energy are exothermic reactions. Some reactions, called endothermic, do the opposite. To say that a reaction "consumes energy" usually refers to transforming potential energy into motion.

Exothermic reactions occur readily, and once they begin most do not need any outside help to carry on. This is similar to the process of diffusion (*see* pp. 17–18), which takes place as molecules spontaneously move from high to low concentrations. An example of an exothermic reaction is the burning of gasoline in the presence of oxygen.

Endothermic reactions generally require an "input" of thermal or kinetic energy to proceed. These reactions transform thermal energy into potential energy. Thus, energy is stored in the chemicals, and their bonds, that are made by the reaction. An example of an endothermic reaction is the conversion of two molecules of ammonia (NH_3) into N_2 and $3H_2$.

Some reactions in cells are exothermic, and some are endothermic. The way cells manage their energy "budget" is to use a go-between—a molecule generated by

Key Points

There are many ways that cells use energy. Here are a few examples:

1. active transport–moving materials across the cell membrane against the natural flow of diffusion;
2. muscle contraction;
3. brain cells processing information (the brain is a greedy consumer of energy–it has only 2 percent of the body's weight but uses 20 percent of its energy);
4. synthesizing and maintaining vital molecules such as DNA; and
5. repairing tissues.

▼ *Racing cars get the energy to move fast from exothermic reactions–gasoline burning in air–that occur in their engines.*

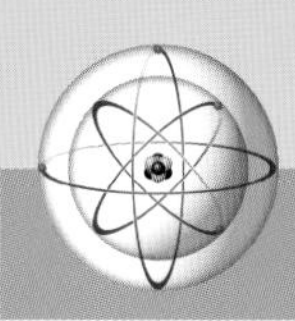

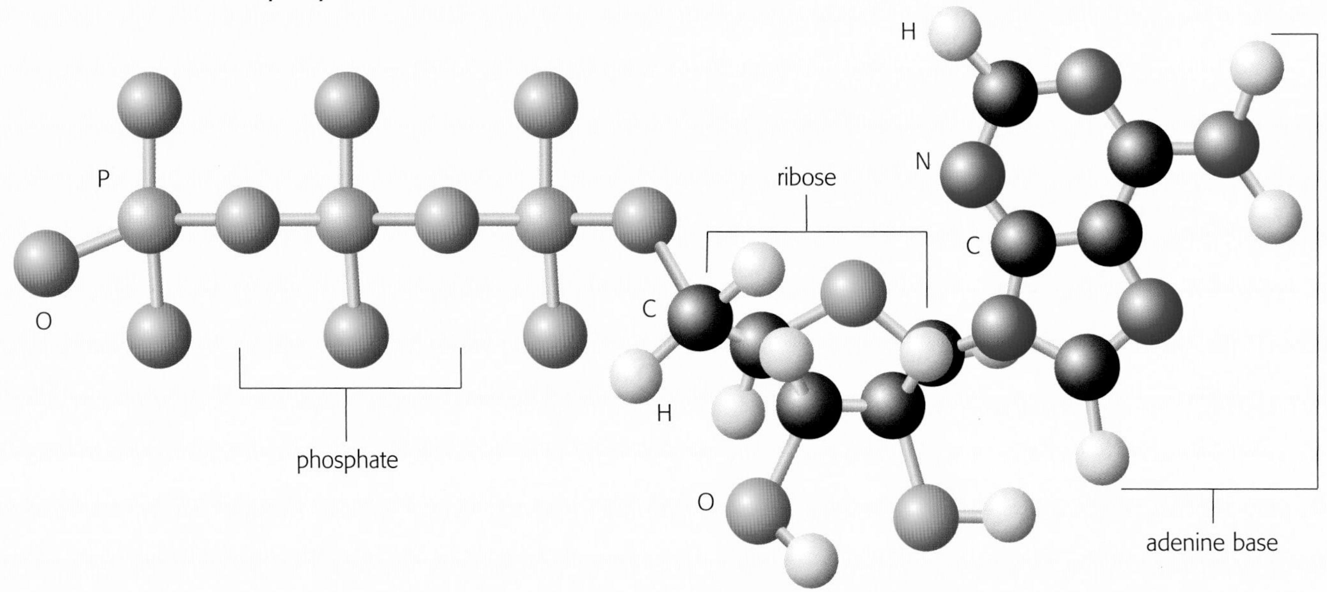

exothermic reactions and consumed by endothermic reactions. The molecule that acts as the go-between is called adenosine triphosphate (ATP).

ADENOSINE TRIPHOSPHATE

ATP consists of the nucleotide adenine, ribose, and a chain of three phosphates. Cells must maintain adequate levels of this molecule to survive. The cell makes ATP by adding a phosphate group to adenosine diphosphate (ADP). This reaction is called phosphorylation.

ATP is often considered the cell's energy currency. ATP is a medium of energy exchange, like money to be exchanged for goods and services. The bonds holding the phosphate groups on the end of the molecule contain potential energy. When endothermic reactions are needed, ATP molecules become involved. The ATP molecules are broken down, and the potential energy stored in their bonds is released.

With the aid of ATP, biochemical reactions can occur as their products become needed. However, providing energy alone is not sufficient to support life. The chemical reactions of cells must take place at the required speed. Thus, cells require enzymes to catalyze (speed up) reactions, controlling the rate at which the reactions generate their products.

ENZYMES

Most enzymes are globular (globelike), water-soluble proteins that either float around or are attached to some part of the cell. An enzyme participates in a reaction, making it go faster, but the reaction does not change the enzyme. The enzyme takes part in the reaction many times and chemically remains the same throughout.

▲ *An adenosine triphosphate molecule is made up of three phosphates, the sugar ribose, and the nucleotide adenine.*

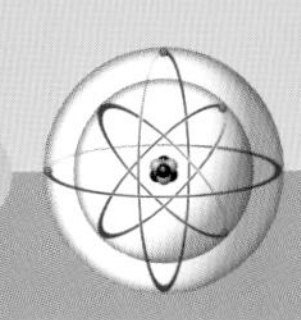

In almost all cases an enzyme catalyzes only one specific reaction. This feature of enzymes is called specificity. The enzyme binds a substrate—one of the reactants that is changed in the reaction—and holds it in the proper place and position. Enzymes are often named after the reaction they catalyze or the reactants they bind. The name ends in the suffix *–ase*. For example, the enzyme lactase catalyzes the breakdown of the sugar lactose, which occurs in milk.

The region of the enzyme that binds to the substrate is called the active site. The binding is temporary, and the enzyme releases the substrate after the reaction is over. The substrate fits snugly into the active site, as a key into a lock, which is why enzymes are so specific—there is only one key. The reaction proceeds quickly because the enzymes hold the reactants in place. The reaction could occur without enzymes but would proceed only slowly, since it would occur only when the reactants bumped into each other by chance.

TRY THIS

Enzymes and apples

Ripe apples contain enzymes, including one called polyphenoloxidase. This enzyme catalyzes the reaction of oxygen with some of the compounds in the apple. The reaction produces dark products and makes the apple turn brown. Like all reactions, including those catalyzed by enzymes, temperature affects the rate at which the reaction occurs.

1. Cut an apple in half; place one half in a refrigerator and leave one uncovered at room temperature.
2. Examine the apple halves every 20 minutes for two hours. Note the rates of the reaction under the two different conditions by observing the color of the apple.

The enzyme lactase is often used in the manufacture of ice cream to give a sweeter taste and a smoother texture. Ice cream without lactase feels "sandy."

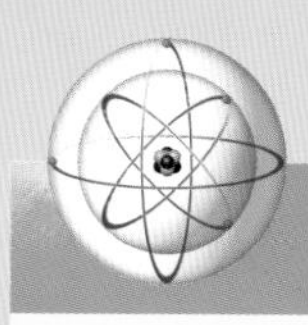

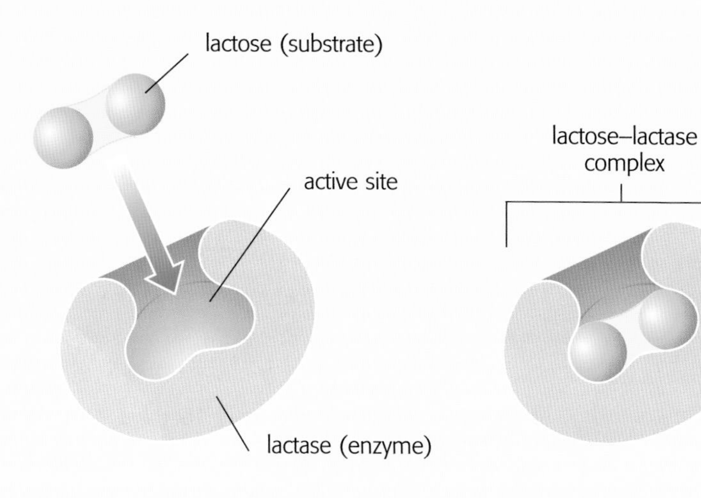

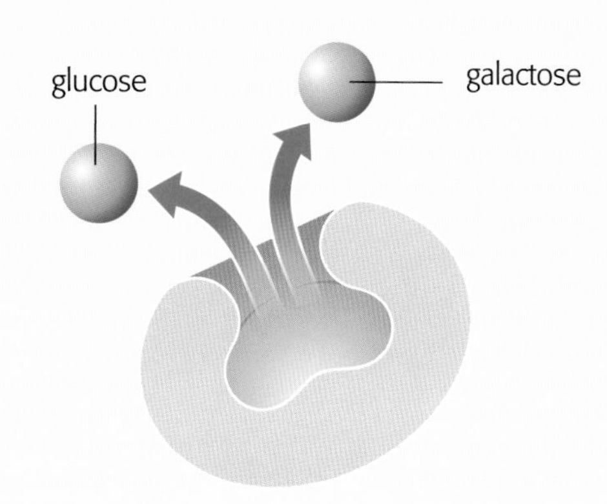

▲ *The enzyme lactase is used in ice cream to make it taste nicer. The lactase binds to the lactose (milk sugar) substrate. The reaction produces sweeter-tasting glucose and galactose.*

Temperature affects the speed of reactions because higher temperatures make the molecules (including enzymes) move faster. With faster motion, enzymes bind more molecules per second.

CARBOHYDRATE CATABOLISM

Enzymes speed up all kinds of reactions in the body. Some of the most important involve carbohydrate catabolism—the breaking down (catabolism) of carbohydrates. This process provides energy by making ATP molecules. Fats and proteins also provide energy, but carbohydrates are the easiest kinds of food for the body to digest. After a meal, the digestive system breaks the bonds of polysaccharides, converting them into molecules of glucose or fructose. Then the process of making ATP begins.

Chemistry in Action

Losing weight the low-carb way

Some people try to lose weight by avoiding foods rich in carbohydrates. This diet forces the body to use its energy stores, which include deposits of fat. However, there are many different kinds of carbohydrates, and the body needs certain complex carbohydrates for reasons other than energy storage. A better way to lose weight is to avoid candy and sweets, which contain huge quantities of simple sugars. These sugars are the fuel on which the body runs, but an excess of fuel simply goes into storage as fat.

GLYCOLYSIS

The first pathway in carbohydrate catabolism is glycolysis. This series of nine different reactions splits a six-carbon glucose molecule into two three-carbon molecules called pyruvate. The process generates two molecules of ATP and two molecules of nicotinamide adenine dinucleotide (NADH). NADH is another biomolecule that effectively stores potential energy in its bonds. The NADH molecules play an important role later, in another pathway.

Pathways such as glycolysis contain many linked steps, all catalyzed by specific enzymes. The products of one reaction become the reactants of the

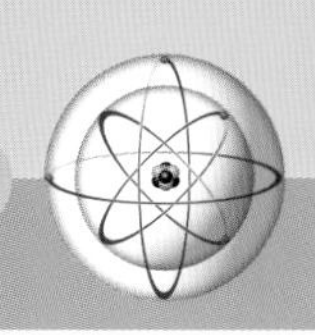

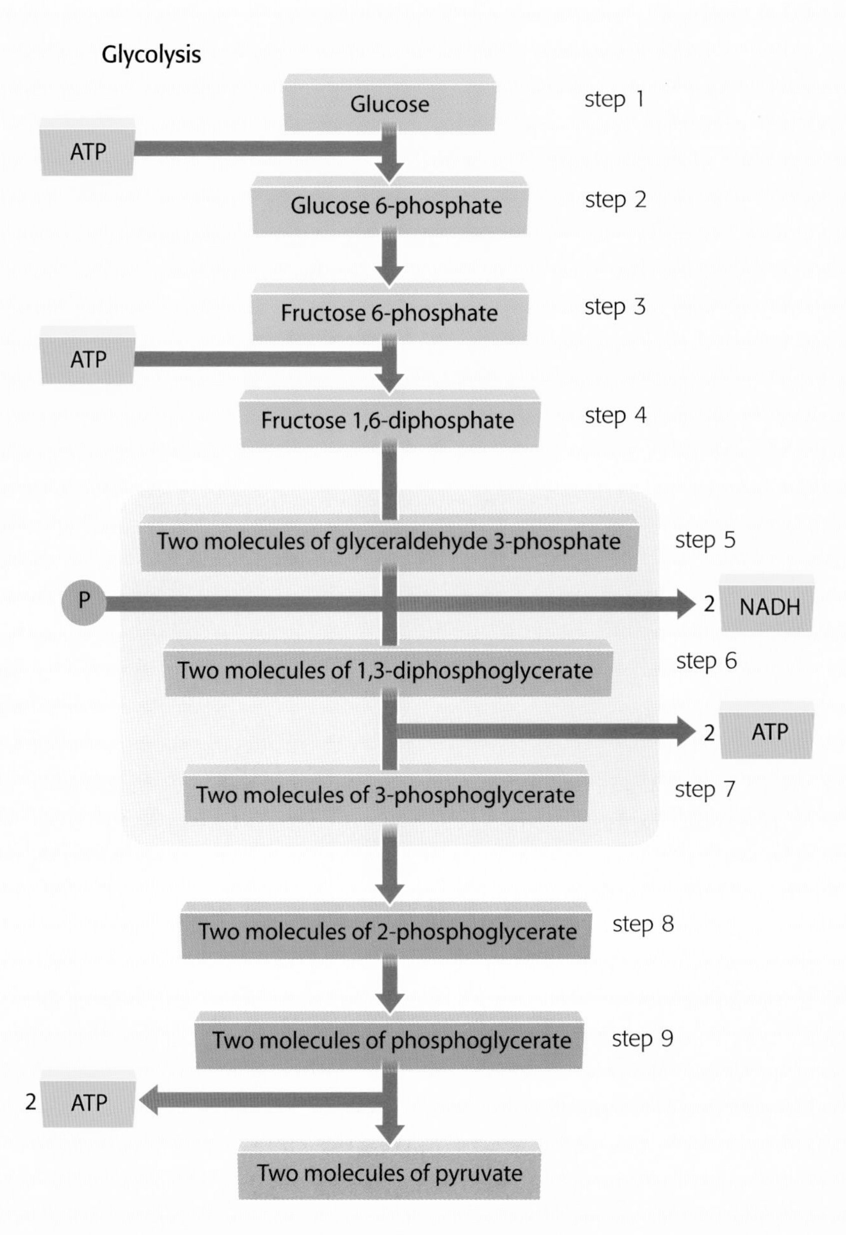

Glycolysis is a series of biochemical reactions in which glucose is broken down to pyruvate. Energy is stored in the form of molecules of adenosine triphosphate (ATP). Energy has to be introduced to the glucose to get the reactions started, and again at step 3; this is called activation energy and is provided by molecules of ATP. Then, a sequence of reactions occurs. The final products of glycolysis are the acid pyruvate and ATP. If glycolysis starts with a single molecule of glucose and two molecules of ATP, the final products are two molecules of pyruvate and eight molecules of ATP.

next. These steps extract the energy of glucose molecules in a slow, controlled process. A rapid, all-at-once reaction, like a fiery explosion, would waste a lot of energy that would end up as thermal energy (heat), making things hotter but not creating any ATP. (Some energy released by carbohydrate catabolism does actually warm the body, helping to maintain body temperature.)

The glycolysis pathway was one of the first catabolism pathways to evolve, and it appeared in living organisms billions of years ago, long before the atmosphere contained much oxygen. (Earth's atmosphere got most of its oxygen later, from the emissions of plants and some kinds of bacteria.) Glycolysis is an anaerobic process—it does not require oxygen. However, glycolysis extracts only about 2 percent of the energy in glucose molecules. This means that there is a great deal of potential energy left in glycolysis products (the two pyruvate molecules).

The energy left over from glycolysis is usually not wasted. Many organisms, including people, extract additional energy from the glycolysis products by aerobic (oxygen-using) pathways. However, certain simple organisms such as bacteria and yeast extract additional energy by an anaerobic process called fermentation. For instance, certain microorganisms extract energy from grains or fruit and produce carbon dioxide and ethanol (alcohol). This is a process that brewers and winemakers have taken advantage of for thousands of years. Fermentation may also occur in our muscle cells when we exercise.

STORING ENERGY

The food we eat supplies glucose, the starting point of glycolysis. Our cells are constantly in need of energy, especially at times of high activity, but we do not have to eat constantly. The body maintains glucose levels by several methods. One way of raising the glucose

▶ *Brewers rely on fermentation to make beer. Yeast is added to a mixture of malt, hops, and water, and fermentation produces energy, carbon dioxide, and beer.*

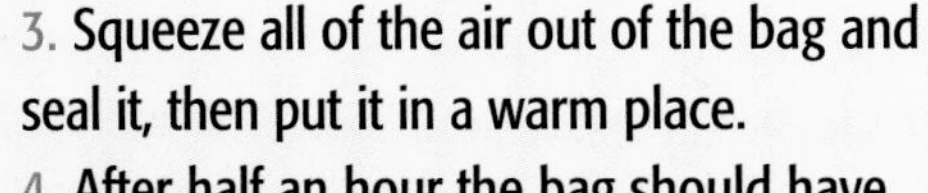

Making gas

You can buy some yeast at the grocery store and carry out your own fermentation experiments.
Materials: cookies, yeast, warm water, Ziploc bag

1. Crush two or three cookies and place the pieces in a sealable bag.
2. Add 1 tablespoon of yeast and about ¼ cup of warm water to the bag. Mix the contents by shaking.
3. Squeeze all of the air out of the bag and seal it, then put it in a warm place.
4. After half an hour the bag should have swollen. The gas produced is carbon dioxide–one of the products of fermentation. You can also try this experiment with sugar, flour, baked beans, and cereal to see which produces the most gas.

▶ *Add warm water and yeast to a bag containing crushed cookies.*

▼ *Trying the experiment with different substances shows that some make more gas than others. This happens because these substances contain more sugar than others.*

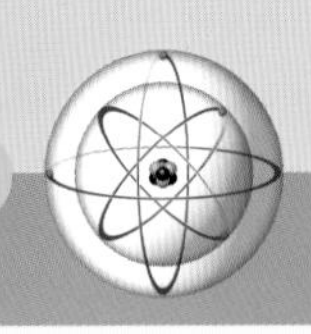

level is by breaking down glycogen, which is a storage molecule that contains many glucose molecules bonded together. Another way to obtain glucose is through gluconeogenesis. This pathway produces glucose by starting with a molecule called lactate or, when we have not eaten in a long time, amino acids or glycerol. Gluconeogenesis occurs in the liver and is triggered by falling levels of glucose in the blood.

MITOCHONDRIA

Most cells in the bodies of people and animals contain an aerobic pathway for squeezing more energy from the products of glycolysis. These reactions occur in structures contained within cells called mitochondria. Mitochondria have an inner and outer membrane and are shaped like tubes or cylinders. Scientists believe that mitochondria were once

▼ *Energy production within a cell is called cellular respiration. It takes place within the cell's mitochondria. Mitochondria are enclosed by two membranes and are usually sausage shaped. When a molecule of pyruvate enters a mitochondrion, a sequence of reactions takes place in the matrix. These reactions produce ATP.*

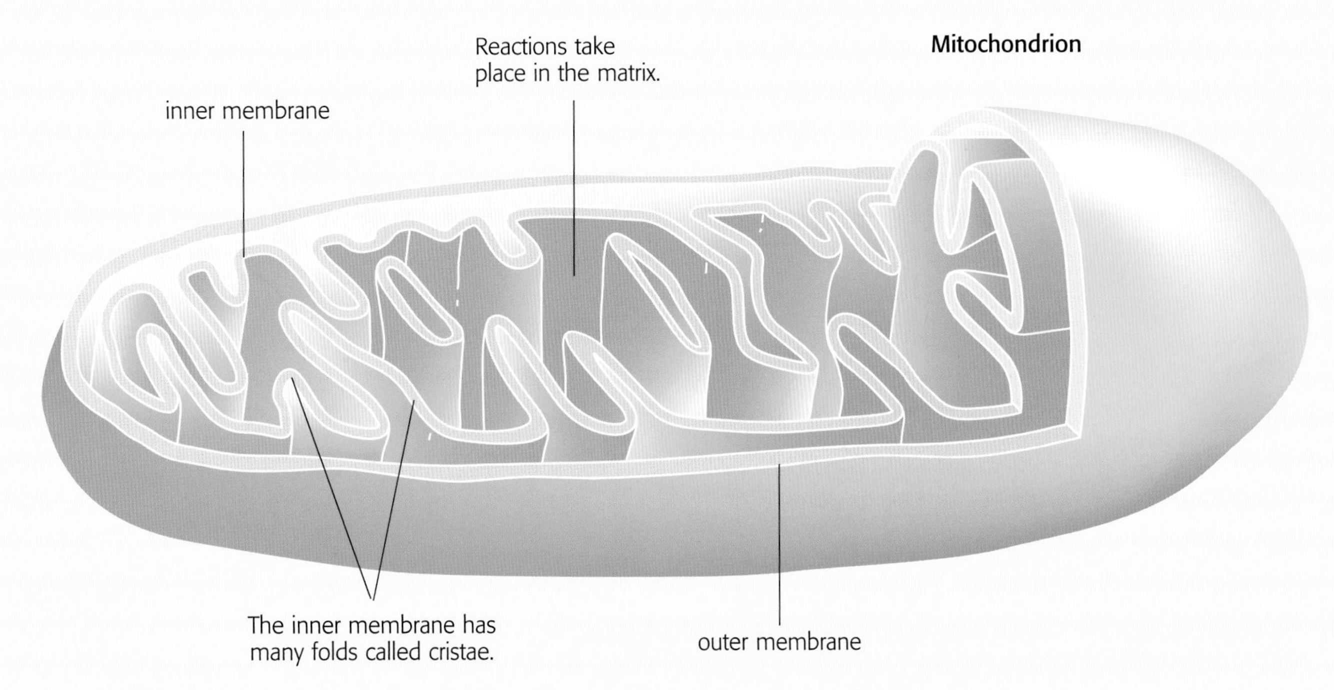

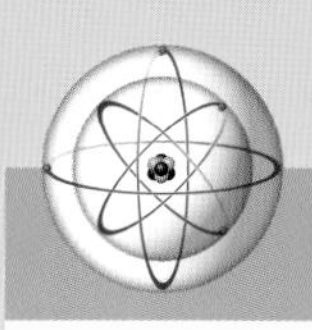

Tools and Techniques

Super-magnification

The details of mitochondria are too small to be seen in microscopes that use light. Instead, scientists use electron microscopes. An electron microscope shoots electrons at targets, such as cellular structures like mitochondria. Although electrons are particles, they also have wavelike properties similar to light. An electron microscope creates an image of the object by detecting the patterns electrons make as they pass through thin slices of the object or bounce off thicker slices.

This electron microscope image of a mitochondrion shows the smooth outer membrane and the folded inner membrane. It has been magnified approximately x50,000.

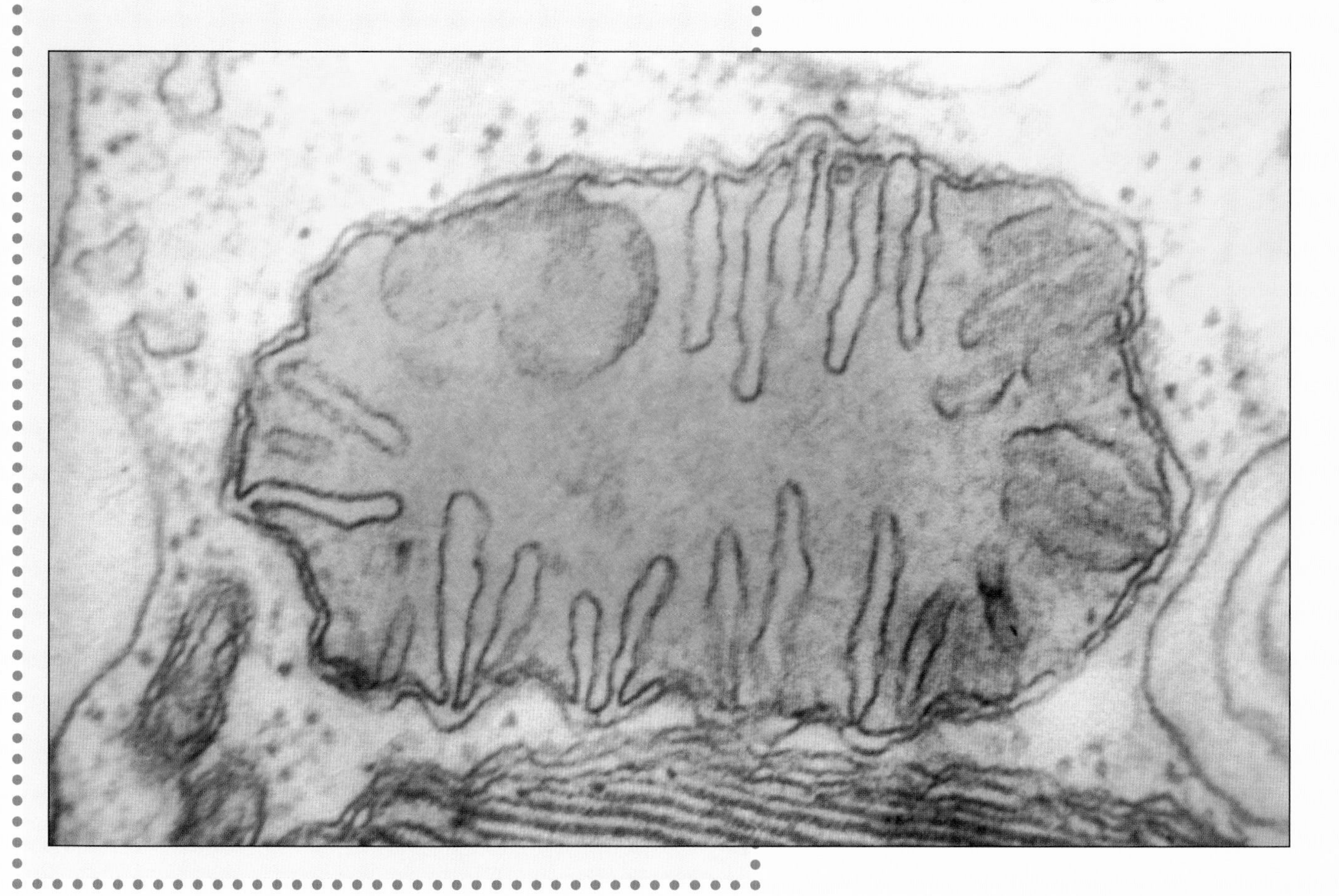

free-living bacteria; hundreds of millions of years ago, one of the cells of an organism captured these bacteria, and over time the two evolved into the useful relationship that exists today.

CITRIC ACID CYCLE

The pathway for aerobic metabolism is called the citric acid cycle. This series of reactions takes the slightly modified products of glycolysis and extracts even more energy, using oxygen and releasing carbon dioxide. These reactions are the reason we breathe in oxygen and exhale carbon dioxide.

The transition between glycolysis and the citric acid cycle occurs when pyruvate, the product of glycolysis,

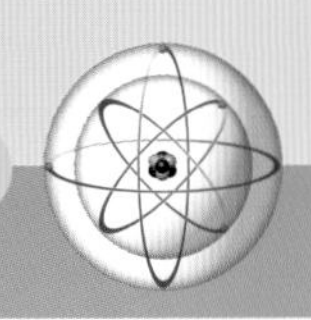

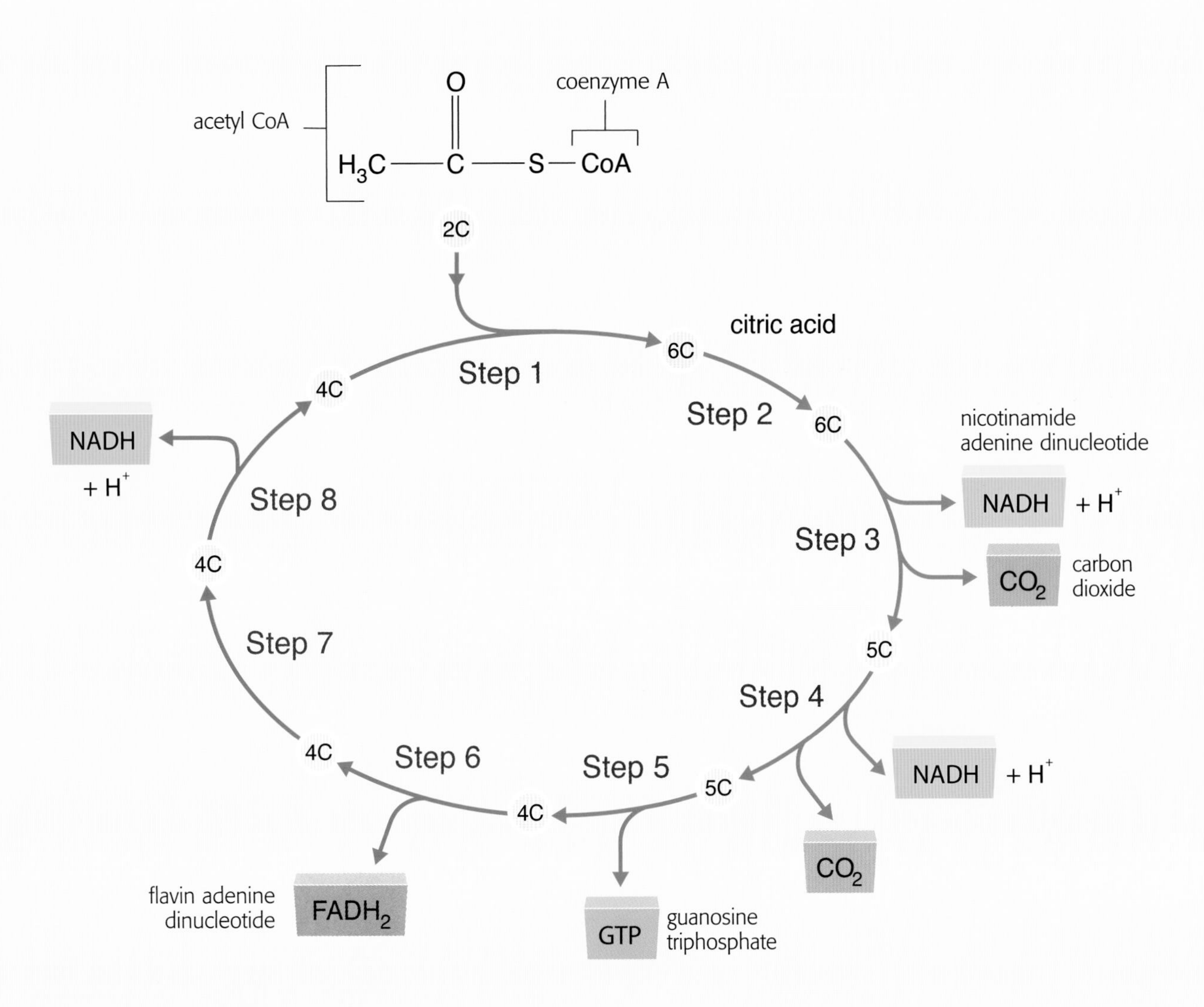

undergoes a reaction. The reaction turns pyruvate into an acetyl group (–$COCH_3$) that binds to a molecule called coenzyme A (CoA). The result is acetyl CoA, the starting point of the citric acid cycle.

The citric acid cycle is cyclical because its starting point—acetyl CoA—is also the end point. The pathway is a loop consisting of eight reactions (steps), each catalyzed by an enzyme. The "citric acid" portion of the name comes from the product of the first reaction in the cycle, when acetyl CoA is converted to citric acid. Another name for this pathway is the Krebs cycle, named for the scientist who first figured out its steps.

Only one of the reactions of the citric acid cycle makes ATP, and it produces only a single molecule of ATP. Other products of the cycle are three molecules of nicotinamide adenine dinucleotide (NADH) and a molecule called flavin adenine dinucleotide ($FADH_2$). These molecules are critical in the next step, in which they lose electrons (they are oxidized). Then, the ADP is phosphorylated (another phosphate group is added), and ATP is produced.

▲ *The citric acid cycle is a loop consisting of eight reactions (steps). Each reaction is made possible by an enzyme. The products of one "turn" of the cycle are: two molecules of CO_2; three molecules of NADH; one molecule of GTP; and one molecule of $FADH_2$.*

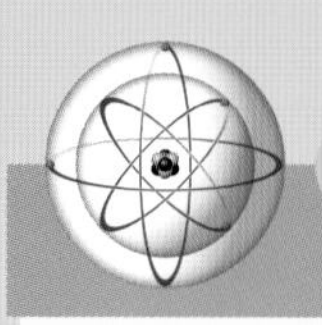

Profile

Hans Krebs (1900–1981)

Hans Krebs (1900–1981) was born in Germany but moved to England in 1933 after the Nazi government ended his job. In the late 1930s Krebs used his expert knowledge of enzymes and chemical reactions to discover the details of the citric acid, or Krebs, cycle. In 1953 he won a share of the Nobel Prize for Physiology or Medicine for his outstanding contributions to biochemistry. He was knighted (giving him the title "Sir") in 1958.

Hans Krebs was awarded his Nobel Prize in 1953, while he was conducting research at Sheffield University, England.

ELECTRON TRANSPORT

NADH and $FADH_2$ are reducing agents, which means they donate electrons in chemical reactions. They give these electrons, originally from glucose molecules, to acceptors (oxidizing agents). In the process, NADH and $FADH_2$ are oxidized, becoming NAD^+ and FAD, respectively. (NAD^+ and FAD take part in the citric acid cycle, where they are reduced again to NADH and $FADH_2$, respectively.) Molecules embedded in the inner membrane of mitochondria pass the electrons along a chain, called the electron transport system.

The electron transport system (called oxidative phosphorylation) takes energy from the electrons as they move along the chain. The energy gained by this process pumps hydrogen ions across the inner membrane. As the hydrogen ions diffuse back, their movement drives an enzyme called ATP synthase. This adds a phosphate group to the ADP to generate ATP. The electron transport system normally yields another 32 molecules of ATP per glucose molecule. The total production of ATP from glycolysis, the citric acid cycle, and the electron transport system is on average 36–38 molecules of ATP per initial glucose molecule. That is about 40 percent of the available energy.

FATTY ACID METABOLISM

Fats are a rich source of energy, providing more than twice as much energy as the same weight of carbohydrate. Some of the energy derived from food is not immediately needed, and the body stores this energy as triglycerides in adipocytes. Triglycerides are "burned" as needed, particularly during exercise.

Fatty acids are broken down into acetyl CoA by a series of enzyme-catalyzed

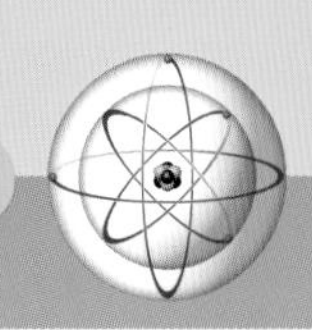

reactions. The fatty acids then enter the citric acid cycle, where they eventually produce ATP. The process begins when triglyceride molecules are split in two by reacting with a molecule of water. This process is called hydrolysis and undoes the condensation reactions (which produce a molecule of water) by which fatty acids attach to glycerol to make triglycerides. Triglyceride hydrolysis produces free fatty acids, which are transported to mitochondria, where further reactions generate acetyl CoA. Triglycerides produce much energy; for example, a molecule of a fatty acid called palmitic acid yields 129 molecules of ATP.

Key Terms

- **Carbohydrate:** One of a group of compounds that includes sugars, starch, and cellulose. Some are essential in turning food to energy. Others are energy stores in plants, and still more act as supporting structural material in plants and animals.
- **Enzyme:** A protein that speeds up chemical reactions in organisms.
- **Fatty acid:** A compound made up of an unbranched hydrocarbon chain with an even number of C atoms and a carboxyl group.
- **Hydrolysis:** The process by which a molecule splits after reacting with a molecule of water.
- **Triglyceride:** A major component of fats. They are a concentrated food energy store in organisms.

Chemistry in Action

Fat and health problems

Adipocytes are cells that have much of their volume taken up by fat (*see* p. 14). Fat adds a lot of weight to the body and can make exercise harder. Too much stored fat also contributes to health problems such as high blood pressure and heart disease. The only sure way to lose weight is to decrease food intake and to exercise more. A lower level of food intake forces the body to use up stored energy sources, and the body increases its use of triglycerides during times of heavy activities such as exercise.

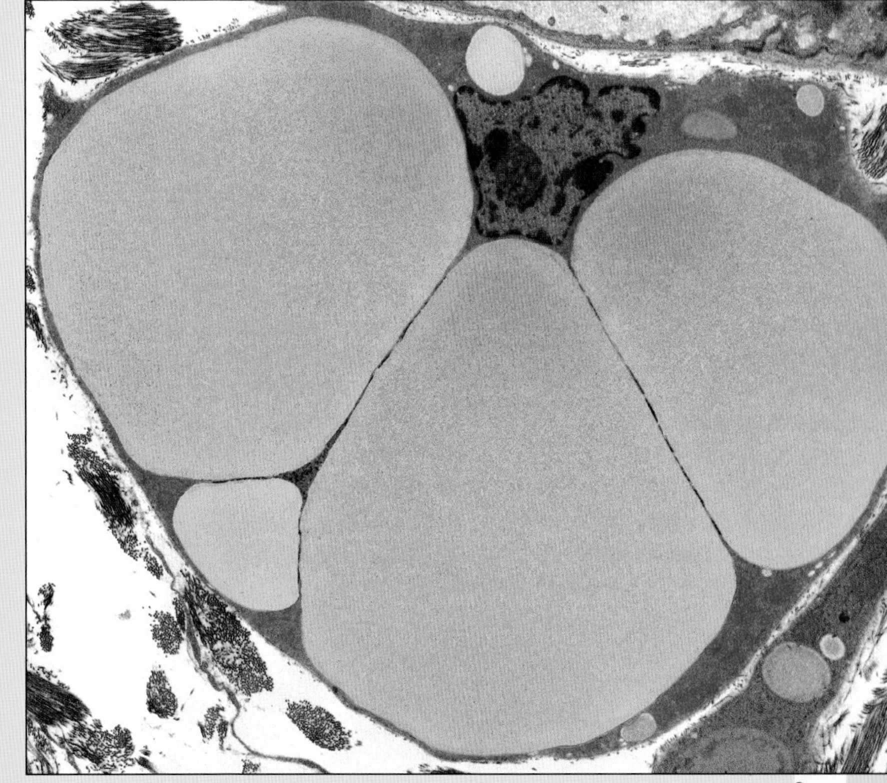

▶ *The yellow areas of this adipocyte cell are deposits of fatty triglycerides. The purple region is the cell's nucleus.*

▲ This cow is starving. If an animal does not eat enough food, its fat deposits will be used up faster than they are replaced. When its fat has been used in metabolism, the animal's muscles will be broken down for energy. The animal will become weak and eventually die.

Acetyl CoA can also be the starting point to make triglycerides. The body converts excess acetyl CoA, produced by an overabundance of carbohydrates, into triglycerides to be stored for later use.

AMINO ACID METABOLISM

The citric acid cycle is important not just in breaking down (catabolizing) molecules and extracting energy, but also in anabolic (building up) reactions. The citric acid cycle is involved in building up as well as breaking down amino acids.

The primary function of amino acids is to make proteins. Excess amino acids from the diet are not stored by the body, and a portion of these molecules can be broken down into products that take part in the citric acid cycle. The amino acids are oxidized and used as an energy source. After a prolonged time without food, the body will begin to break down its own proteins and convert the amino acids into energy in a desperate effort to sustain life. The opposite reactions can also occur. Then, enzymes catalyze reactions in which molecules involved in the citric acid cycle and other pathways are made into amino acids. For example, pyruvate can be used to make the amino acid alanine.

NUCLEOTIDE METABOLISM

Nucleotides are components of many important substances of the body, such as the nucleic acids RNA and DNA (*see* pp. 26–27). Unlike many of the molecules of life, however, we do not get many of these vital nucleotides from our diet. Instead, several amino acid pathways are involved in making nucleotides. For example, the amino acids aspartate, glutamate, and glycine, along with other molecules, are used to make nucleotides. The body also salvages nucleotides when RNA and DNA molecules break down.

Key Terms

- **Amino acid:** A compound made up of a carboxyl group (–COOH) and an amino group (–NH_2) attached to the same C.
- **ATP:** A nucleotide that carries chemical energy in living organisms. It is made up of an adenine base, ribose, and three phosphate groups.
- **Nucleotide:** A molecule made up of a nitrogen-containing base, a sugar (such as ribose), and a phosphate group.

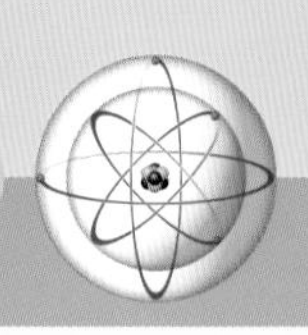

The catabolic reactions concerning nucleotides are different from those involving carbohydrates, fats, and proteins. Nucleotides are not very important sources of energy. They are broken down into various molecules to be used in other pathways or are excreted (removed from the body) in the urine.

PHOTOSYNTHESIS

The reactions that break down glucose and generate ATP liberate a great deal of potential energy (stored energy). Energy cannot be created or destroyed, only changed from one form to another. So it must take a great deal of energy to make glucose molecules. The energy to do this comes from sunlight.

▼ Sunflowers and other green plants use the energy of sunlight to make carbohydrates from carbon dioxide and water. This process is called photosynthesis, and it also makes the oxygen that animals need to live.

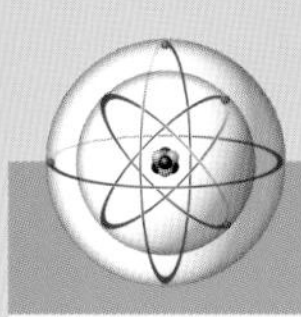

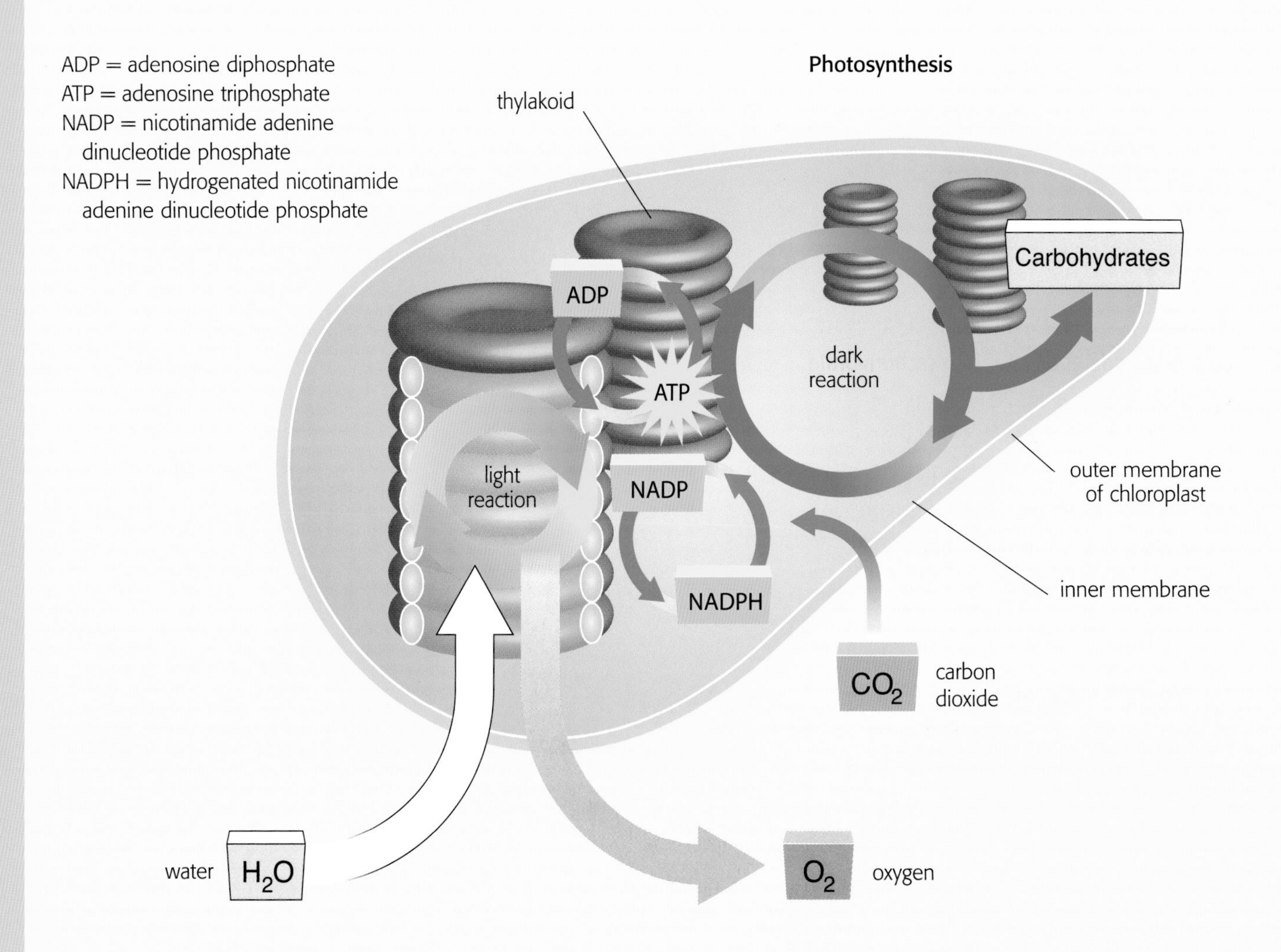

▲ *Photosynthesis takes place in chloroplasts, which are tiny organs within plant cells. There, chlorophyll absorbs sunlight, and light reactions produce ATP and NADPH. Elsewhere in the chloroplast, a series of dark reactions fix carbon dioxide from the air and produce carbohydrates.*

Photosynthesis is a metabolic pathway in green plants, algae, and a few bacteria and other single-celled organisms. The word *photo* means "light," and *synthesis* means that the light energy is used to make, or synthesize, substances. Photosynthesis makes carbohydrates and ultimately the food that fuels all living organisms on the planet.

Photosynthesis begins with light-absorbing molecules called chlorophyll. Other light-absorbing molecules are also involved. Chlorophyll and its associated molecules occur inside plant cells inside tiny organs called chloroplasts. Many of a plant's cells have these structures, but they are especially active in leaves.

Plants use sunlight to make ATP and electron-carrier molecules called nicotinamide adenine dinucleotide phosphate (NADPH). Then comes a series of reactions called dark reactions (because they do not involve light). These reactions involve a cycle called the Calvin cycle (named for its discoverer, American chemist Melvin Calvin, 1911–1997). The reactions store the chemical energy in a more stable,

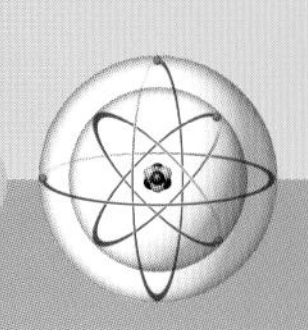

A Closer LOOK

Photosynthesis

The equation for photosynthesis can be written as:

$$6CO_2 + 6H_2O + \text{light} \rightarrow C_6H_{12}O_6 + 6O_2$$

Photosynthesis uses light energy to convert carbon dioxide and water into carbohydrate (glucose) and oxygen. Plants carry out photosynthesis to make their own food (glucose). Compare this reaction to respiration, the process by which animals derive energy from the oxidation of carbohydrate:

$$C_6H_{12}O_6 + 6O_2 \rightarrow 6CO_2 + 6H_2O$$

Plants harness the energy of sunlight and transform it into chemical potential energy and oxygen. Animals obtain energy from the reverse process, breaking apart the bonds of carbohydrates. This produces energy along with "waste" products of carbon dioxide and water.

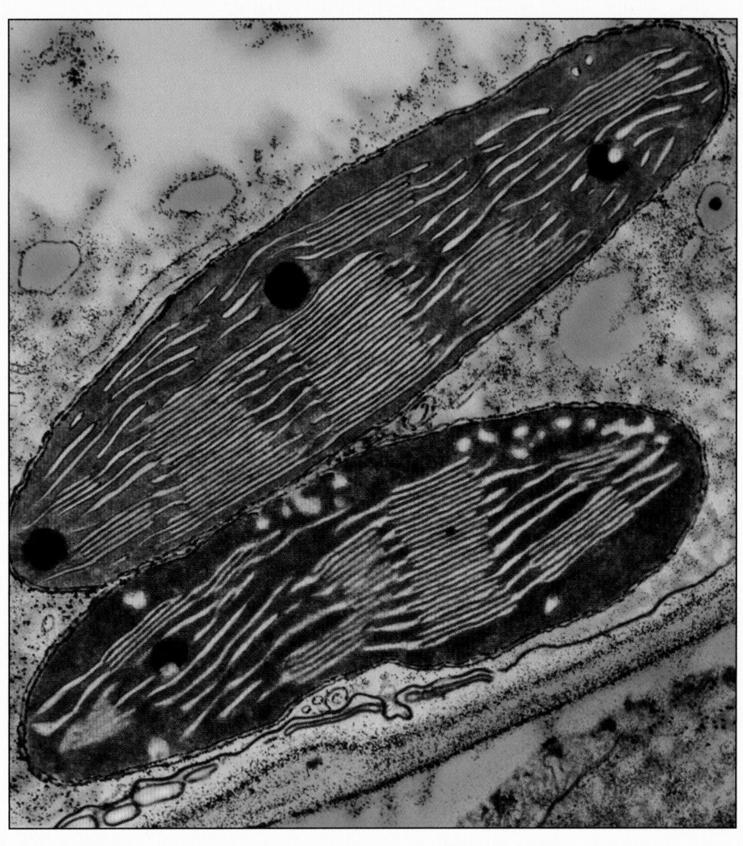

▲ *Two chloroplasts within a plant cell. The light-absorbing system of the chloroplasts is contained in thylakoids, the long, thin, pale structures.*

long-lasting form of carbohydrate molecules for future use. The process fixes carbon dioxide taken from the atmosphere, incorporating the carbon into organic compounds and then on into carbohydrates.

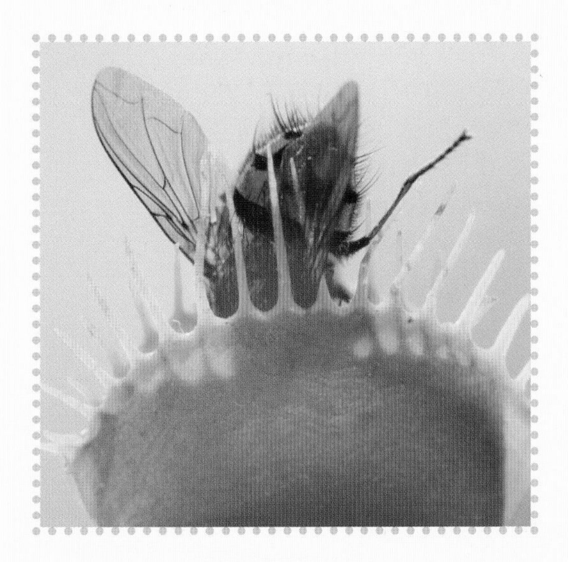

► *Plants need nitrogen to make proteins. Most plants get nitrogen from the soil in which they grow. However, since Venus flytraps live in bogs, where the soil has little nitrogen, they do things differently. They get their nitrogen directly from insects that they catch and digest in their leaves.*

See ALSO ...
- *Lipids, Vol. 9: pp. 12–19.*
- *Proteins and Nucleic Acids, Vol. 9: pp. 20–29.*

5 Making Molecules

Chemical reactions occurring in or around cells synthesize (make) most of the critical molecules of life, including molecules required for storing and interpreting genetic (inherited) information.

Before the 19th century many people believed that chemical reactions in the body and other living things were somehow different from the reactions that take place in the laboratory of chemists. Some people thought there was a special force—a "vital" force—involved in the making of biochemical molecules that was only present in living tissue. If so, there would be no way to synthesize biochemical molecules outside of the body's tissues. Then in 1828, a German chemist, Friedrich Wöhler (1800–1882), synthesized urea, a product of protein metabolism, which is excreted in the urine. Shortly afterward chemists began synthesizing many other organic compounds in their laboratories, proving that no vital force was required for any of these compounds.

How do new bean plants know how to grow? The answer lies in the bean seed, which contains all the information and instructions for making the chemicals that make up a bean plant.

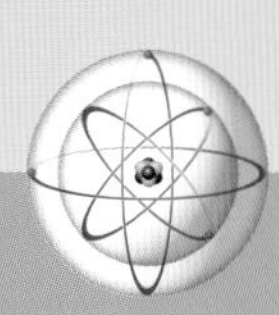

▲ *A child enjoys a glass of orange juice. Humans need fruit and vegetables in their diet because they contain large amounts of vitamin C. The human body is unable to synthesize this vitamin and must obtain it from food.*

ENZYMES

Enzymes are proteins that are produced by the body to catalyze (speed up) synthesizing reactions and other biological reactions, otherwise these processes would occur too slowly. The ability to synthesize a molecule requires two factors:

1. the presence of the substances needed to make the molecule
2. the specific enzyme to catalyze the reaction.

Since enzymes often catalyze only one specific reaction, the synthesis of each molecule needs its own enzyme. If the enzyme is missing, the molecule cannot be synthesized by the body and must be obtained from sources such as food. For example, most animals can synthesize vitamin C (ascorbic acid). However, humans cannot synthesize vitamin C because we lack the necessary enzyme. For us, vitamin C is a nutrient we must obtain from eating oranges,

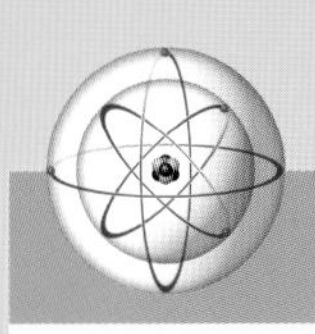

Profile

Friedrich Wöhler

Friedrich Wöhler (1800–1882) was a German chemist and professor at the University of Göttingen in Sweden from 1836 until his death. He was a devoted teacher, often starting his classes very early in the morning, and wrote several chemistry textbooks. By making the compound urea from an inorganic salt, ammonium cyanate, Wöhler helped show that the principles governing biochemical molecules are the same as those for other substances. He also isolated the elements aluminum and silicon.

Friedrich Wöhler proved that biological compounds could be made in the laboratory.

lemons, green leafy vegetables, or other foods rich in this vitamin.

Enzymes not only speed up reactions, they are also part of the body's way of controlling the reactions. In synthesis reactions, the body must make enough of the needed molecule but not too much. To control reactions, the body often regulates the activity of an enzyme. Sometimes a synthesized molecule itself acts to regulate the enzyme, binding to the enzyme and changing its shape, so its active site (the part of the enzyme that takes part in reactions, *see* p. 33) is no longer accessible or functioning. When enough copies of the molecule have been synthesized, there are plenty of copies to bind the enzyme, temporarily reducing production until more are needed. This type of regulation mechanism is known as feedback. Because regulation reduces, or inhibits, the rate of the reaction, it is an example of negative, or inhibitory, feedback.

Key Terms

- **Ionized:** Describes a molecule or atom that has acquired a positive or a negative electrical charge through the addition or removal of electrons.
- **Protein:** A molecule made of a chain of amino acids. Proteins are vital compounds for all organisms.
- **Vitamin:** Substance an organism needs to function.

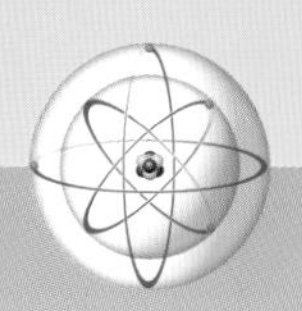

SYNTHESIS OF AMINO ACIDS

Amino acids are the building blocks of proteins. Twenty different amino acids are used to make proteins in most living organisms. For humans, 11 of these amino acids can be synthesized in the body. (Sometimes more or fewer amino acids can be synthesized, because of differences in children and adults and the existence of alternative pathways.)

Nitrogen is an important component of amino acids that people obtain from food. Almost all of this nitrogen originally came from plants. Plants obtain nitrogen from the air—about 80 percent of air is nitrogen. Bacteria living in the soil or attached to the roots of certain plants "fix" nitrogen (incorporate the element into usable compounds). The nitrogen fixation process involves the turning of atmospheric nitrogen, which exists as the molecule N_2, into ammonia (NH_3). This process occurs with the help of a large enzyme called nitrogenase. The ammonia quickly becomes ionized (electrically charged) and plants use this reactive substance in the synthesis of amino acids and other nitrogen-containing molecules. Like photosynthesis and carbohydrate production, nitrogen

▼ *The pink nodules on these pea plant roots are filled with bacteria. The bacteria fix nitrogen from the atmosphere and convert it into an organic form that the plant can use to produce proteins.*

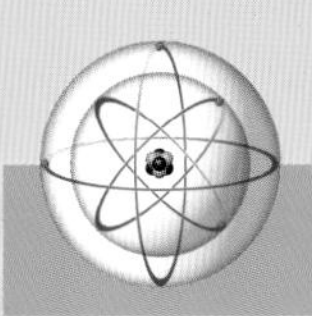

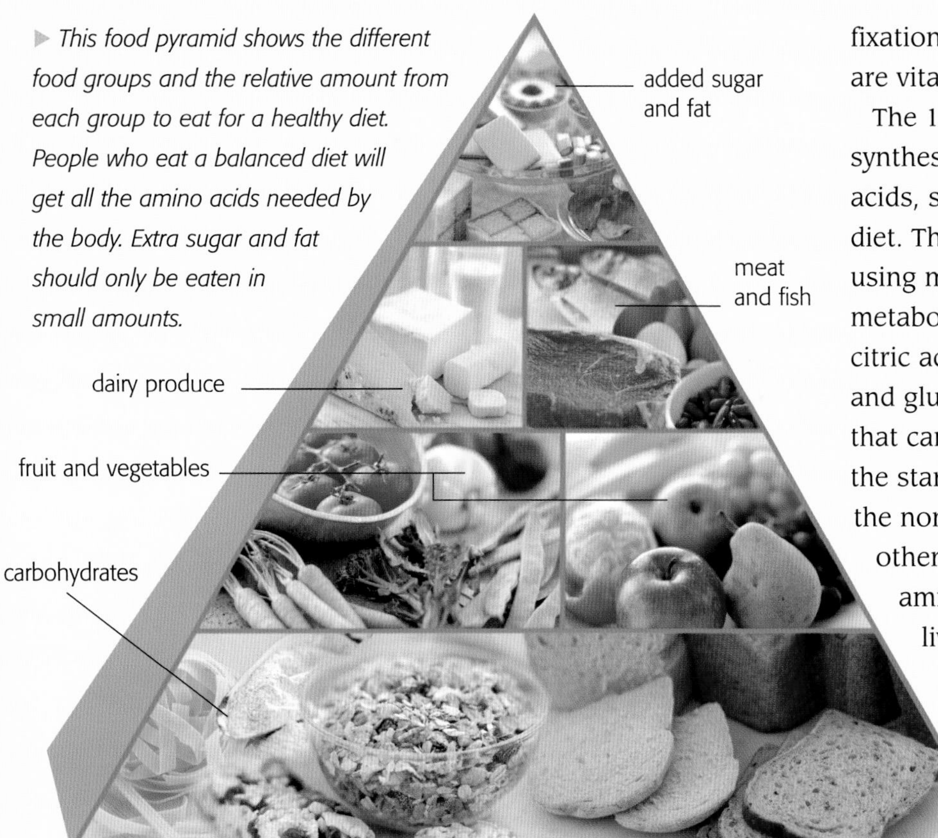

This food pyramid shows the different food groups and the relative amount from each group to eat for a healthy diet. People who eat a balanced diet will get all the amino acids needed by the body. Extra sugar and fat should only be eaten in small amounts.

fixation is another reason why plants are vital to all life on Earth.

The 11 amino acids that the body can synthesize are called nonessential amino acids, since they are not essential in the diet. These amino acids are often made using molecules involved in other metabolic pathways, particularly the citric acid cycle (*see* p. 38). Glutamate and glutamine, two of the amino acids that can be synthesized by humans, are the starting point for making many of the nonessential amino acids as well as other important molecules. These two amino acids are so critical that all living organisms have the enzymes glutamate dehydrogenase and glutamine synthetase, which catalyze the synthesis of glutamate and glutamine, respectively.

A Closer LOOK

Amino acids

In humans, nine of the twenty amino acids used to make proteins must be obtained from food sources. These nine amino acids are called essential amino acids, since they are essential components of a healthy diet. Our bodies acquire these amino acids by digesting proteins from our food. Proteins called complete proteins provide all the essential amino acids. Milk, eggs, and fish are excellent sources of complete proteins. Although most grains and vegetables lack one or more of the essential amino acids, eating a combination of these foods will also provide all the essential amino acids. Children need more histidine and lysine than adults, as both of these essential amino acids help children grow.

Essential amino acids	Nonessential amino acids
Histidine	Alanine
Isoleucine	Arginine
Leucine	Asparagine
Lysine	Aspartic acid
Methionine	Cysteine
Phenylalanine	Glutamate
Threonine	Glutamine
Tryptophan	Glycine
Valine	Proline
	Serine
	Tyrosine

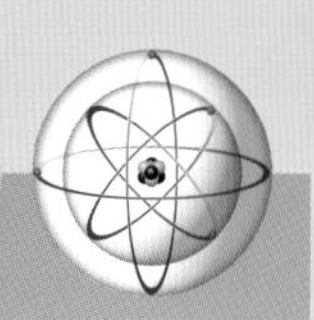

Key Points

The reactions of glycolysis (*see* p. 34) and the citric acid cycle (*see* p. 38) provide pathways that derive chemical energy from breaking down substances like glucose. However, these pathways are also involved in activities that build new substances because the cell uses many of the same reactants and products to synthesize amino acids, lipids, and other biological molecules.

MAKING LIPID MEMBRANES

A cell needs a membrane to keep the cell's interior separated from the solution that surrounds it, as well as to regulate the traffic of molecules into and out of the cell. Cell membranes are largely made of lipids called phospholipids (*see* p. 15). The body must ensure that it has sufficient quantities of phospholipids in order to grow and maintain its tissues. Although membranes are broken down and recycled, synthesis is also important.

The chemical phosphatidate is a simple phosphoglyceride, a common type of phospholipid. Phosphatidate is involved in several pathways that synthesize more complex phosphoglycerides, which are components of cell membranes. Phosphatidate is also often involved in the synthesis of triglycerides (*see* p. 14), the fats in which the body stores energy. Phosphatidate itself comes from a molecule called glycerol 3-phosphate by a series of reactions involving coenzyme A. Coenzymes are small molecules that are needed to help some enzymes work. "A" denotes that this molecule supplies a

Tools and Techniques

Gel electrophoresis

Gel electrophoresis is a technique that scientists use to separate mixtures of proteins or nucleic acids. Proteins and nucleic acids are large, heavy molecules that have side chains of compounds that form ions in solution. Ions have a positive or negative electrical charge because some atoms in the molecule have gained or lost electrons. When an electric current is passed through a solution, ions move toward the oppositely charged electrode.

Several drops of mixed protein solution are applied to the middle of a gel plate that has a positive electrode at one end and a negative electrode at the other. When a current is applied, the protein molecules are either attracted or repelled by the electrodes, depending on their charge. The gel acts like a molecular sieve that separates out the proteins according to how heavy they are. The lightest molecules are able to move through the gel fastest and will get closest to the relevant electrode. After a set period of time, the current is turned off and the plate is rinsed with a staining fluid. The proteins are revealed as a series of bands or spots. Each protein travels a certain distance according to its mass and charge. The spots on the gel can then be sampled and examined chemically to determine which protein is which.

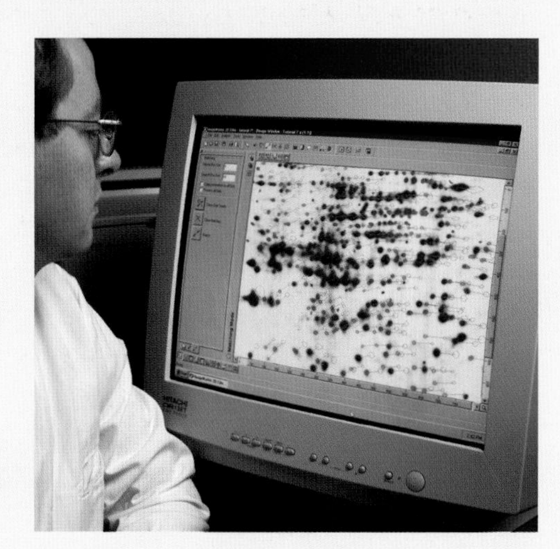

A scientist looks at the results of a gel electrophoresis test on a computer screen. He can work out which proteins are present by their position on the gel.

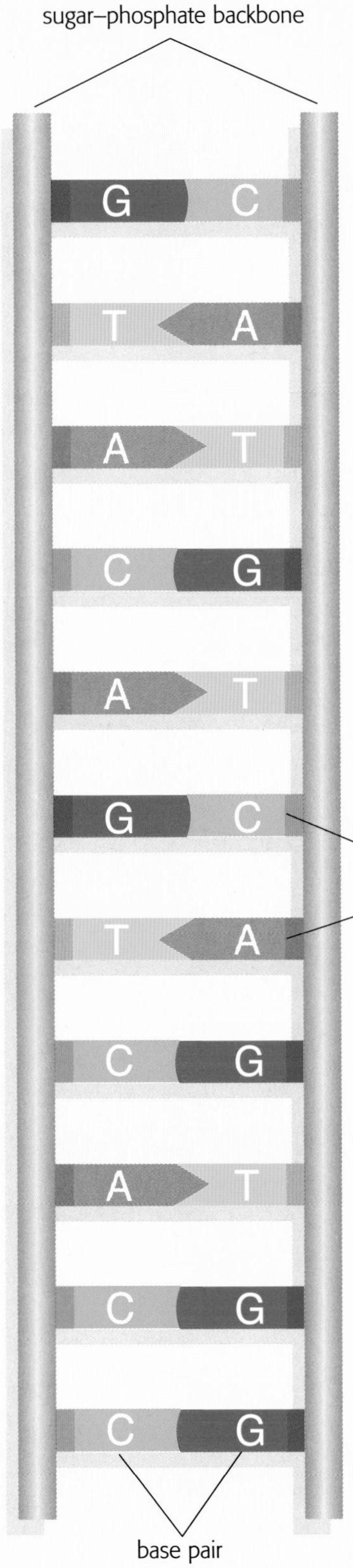

sugar–phosphate backbone

G C
T A
A T
C G
A T
G C
T A
C G
A T
C G
C G

nucleotide bases

base pair

◄ The basic structure of a DNA molecule resembles a ladder. The uprights of the ladder are made of a sugar phosphate. The rungs are nucleotide bases. Each rung has two bases that complement each other. Adenine (A) is always paired with thymine (T) and cytosine (C) is always paired with guanine (G). The ladder is twisted into a spiral double helix.

particular organic group that the enzyme needs to function. Coenzyme A has a key role in the citric acid cycle (*see* p. 38).

SYNTHESIS OF NUCLEOTIDES

Nucleotides are the building blocks of deoxyribonucleic acid (DNA) and ribonucleic acid (RNA). Both of these molecules (*see* p. 27) carry genetic information. Like lipids in cell membranes, old nucleotides are often recycled. When cells disassemble old or unneeded RNA or DNA, the nucleotides usually go into making new molecules of RNA or DNA. Synthesizing nucleotides from their basic components costs the cell more energy than recycling old molecules, but sometimes this is necessary.

GENETIC INFORMATION

The long chains of nucleotides that make up the nucleic acids RNA and DNA are part of the cell's method of storing genetic information and carrying out its instructions. The process of storing,

A Closer LOOK

Going with the flow

Scientists working in genetics and molecular biology have shown that the flow of information for making new proteins always travels in the same direction. The instructions for a new protein start with DNA where the codes for amino acid sequences are transcribed (copied) by RNA. The codes on the RNA are then translated into the amino acids that make up a protein.

DNA —— transcription ——► RNA —— translation ——► protein

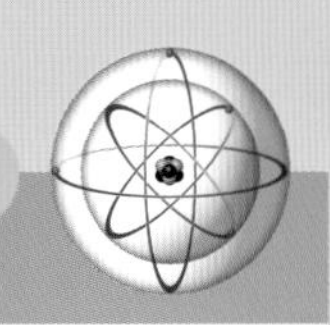

Key Terms

- **Nucleic acid:** A chain of nucleotides.
- **Nucleotide:** A molecule made up of a sugar, a nitrogen-containing base, and a phosphate group.

maintaining, and acting on genetic information involves molecular synthesis on a large scale and in a precise manner. The sequence of nucleotide bases in RNA and DNA contains information that must be preserved during synthesis of these long molecules. The cell uses these sequences as templates to synthesize the many different proteins it needs to carry out its function. The flow of genetic information goes from DNA, which holds the information, to RNA, which transmits it to the enzymes that synthesize proteins.

Most of a human cell's DNA resides in the nucleus, the central structure of a cell. Growth of tissues involves cell division, since cells grow and form new tissue by dividing into two, producing two cells. When a cell divides, its DNA must replicate (copy itself) so that both daughter cells have the same DNA as the parent cell.

DNA REPLICATION

Replicating DNA involves making new DNA. This synthesis does not proceed at random. Because the DNA's sequence contains information, the sequence must

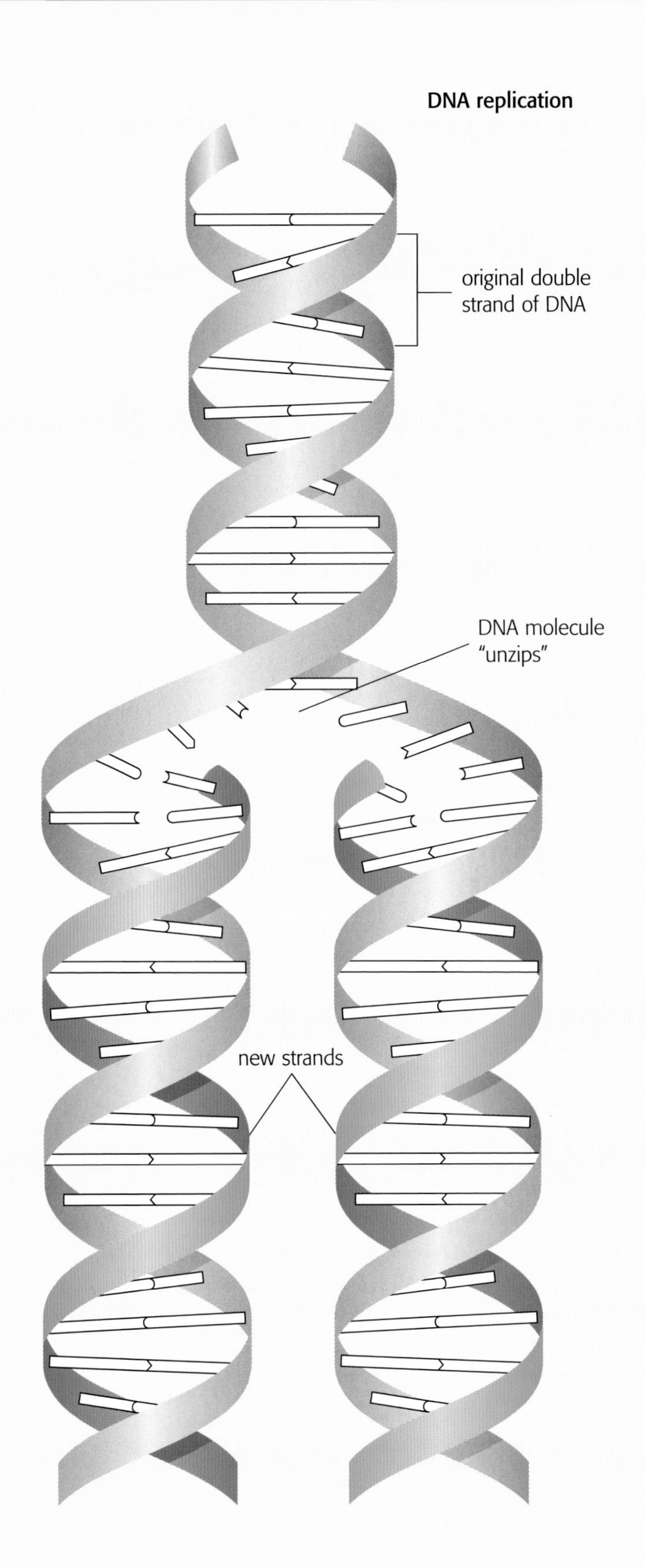

When DNA replicates, the original double strand of the molecule breaks apart so each strand can be copied. The new strands are identical to the original double helix.

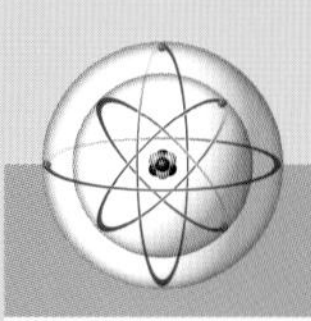

be preserved as much as possible. This need to maintain the sequence is important: the sequence must be copied exactly or the wrong or inactive proteins will be produced.

Prior to cell division, the cell replicates its DNA and forms it into x-shaped structures called chromosomes. Chromosomes are long, coiled strands of DNA that carry genetic information. Replicating the base sequence involves pulling apart the double-helix structure of DNA. That is not difficult because the strands are bonded with weak hydrogen bonds instead of strong covalent ones (*see* vol. 1: pp. 42–55). After they are separated, each strand becomes a template for building its double-helix partner. A large enzyme called DNA polymerase binds to each of the strands and attaches the correct nucleotide, slowly moving down the chain. (The enzyme gets its name because it helps polymerize DNA nucleotides, which means it catalyzes the reaction by which nucleotides link together to form a chain, or polymer.) When the enzyme is finished there are two identical double-helix DNA molecules (if there were no errors in copying).

Each strand of a DNA double helix can serve as the template because the two strands are complementary—the sequence of one strand determines the other. That occcurs because adenine (A) bases always bond to thymine (T) bases, and cytosine (C) bases always bond to guanine (G) bases, and this makes the most stable double-helix structure. If you know the sequence of one strand of a double helix you can always determine the other: the complementary strand of the sequence AAGCAT, for example, is TTCGTA.

Key Terms

- **Base pair:** A base and its partner in a strand of nucleic acid. These are always adenine and thymine (or uracil), and guanine and cytosine.
- **Double helix:** The shape of a DNA molecule. It resembles a twisted ladder.
- **Polymerase:** An enzyme that creates a polymer (long chain) of nucleotides from a template of DNA or RNA.

DNA REPAIR

Most of the time, DNA polymerase attaches the correct nucleotide base. However, there are sometimes errors in copying, so a base pair in a DNA double helix is mismatched: an A is not bonded to a T, or a C is not bonded with

▼ *This frog has three back legs. This mutation has arisen through a mistake when DNA was being copied in the egg. Many mutations, even major ones, have little effect if the DNA sequence is not critical to the organism. In a small number of cases the effect is beneficial to the organism, but the majority of mutations are harmful.*

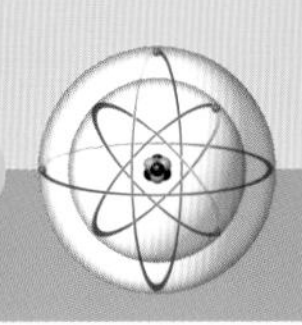

a G. Unintentional changes in the sequence can also occur if the DNA molecule becomes damaged: perhaps a molecule that readily engages in chemical reactions causes a change in one of the nucleotides, or radiation strikes the DNA and breaks the molecule apart. Changes in the DNA sequence are called mutations.

Since the cell relies on information contained in the sequence of DNA, it must try to correct errors. The complementary strands aid error detection because in many situations where there are mismatched pairs, one of the nucleotide bases is correct and the other is an accidental change or a copying error. Figuring out which base is the correct one is not always easy, but sometimes the mistaken base has been chemically altered. Newly copied strands are often chemically tagged, so the cell knows which strand is most likely to contain a copying error. Dozens of enzymes in the cell catalyze actions that detect and correct sequence mistakes. Some, but not all, mistakes are found and corrected.

GENES AND DNA

Extremely important sequences of DNA form genes, a unit of genetic information. Genes control or influence features such as eye color, height, and certain types of behavioral tendencies. A gene does this by coding for a specific

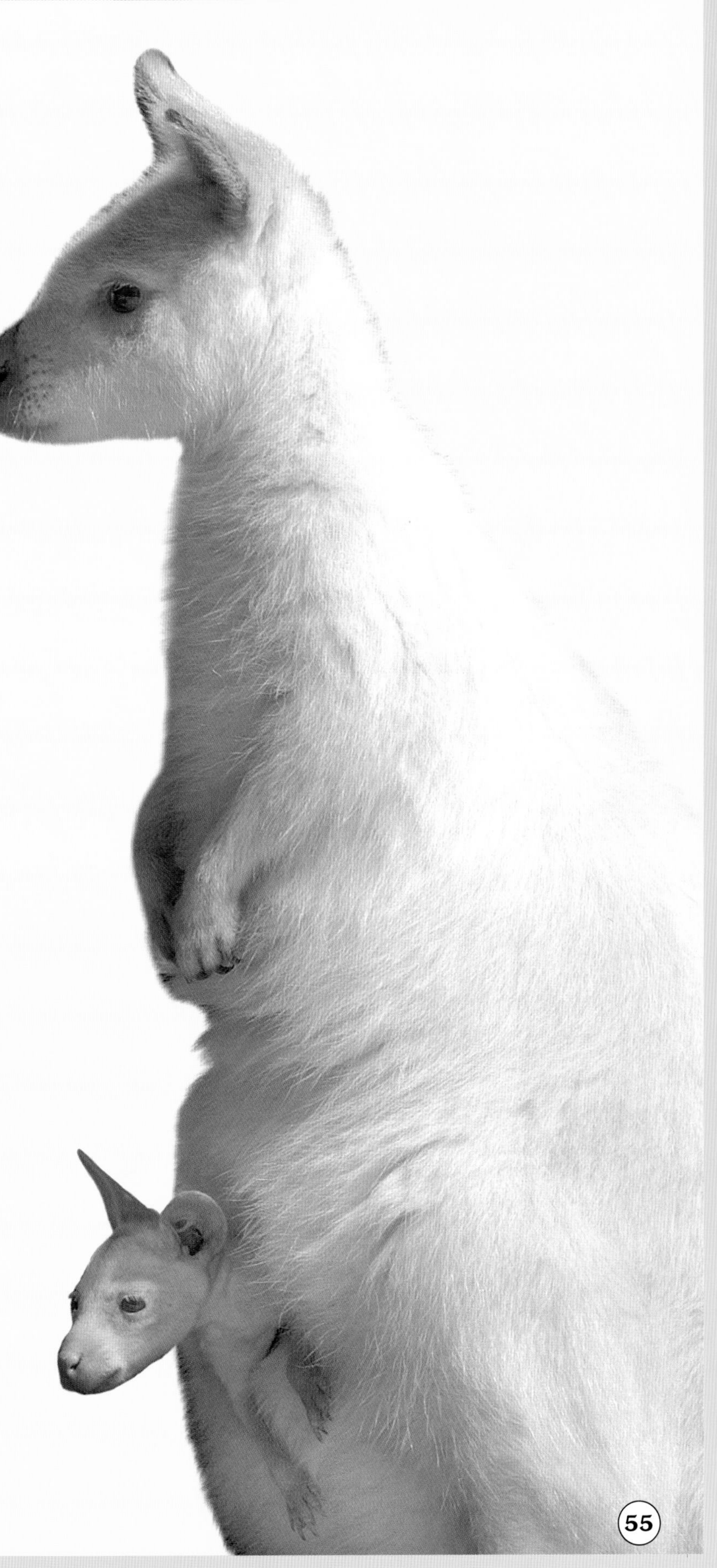

This kangaroo is white and has pink eyes because its genes lack the code for the usual brown kangaroo hair and eye colors. The mother has passed on these features to her baby.

A Closer LOOK

Junk DNA

Only 2 to 4 percent of DNA in humans codes for a gene. Similar percentages code for genes in animals and plants. Some stretches of the DNA strand are highly repetitive, consisting of sequences repeating themselves many times, and frequently have no obvious function. This DNA is sometimes called "junk DNA." Other sequences are involved in controlling the activity of certain genes, regulating how often they are transcribed.

▼ *This electron micrograph shows the process of DNA (pink strand) transcription. During transcription, the enzyme RNA polymerase recognizes a start instruction on the DNA and begins to build a strand of mRNA (green). The RNA polymerase moves along the DNA strand until it reaches a stop code.*

protein. Proteins perform many of the functions of the cell and are also involved in structural properties, so the number and type of these proteins strongly influences how cells behave and interact with one another. The behavior of the cells in turn governs the features of the organism.

Enzymes "read" the sequence of a gene to make the corresponding protein. First, a group of molecules in the cell copies the sequence of DNA, which resides in the nucleus. Special RNA molecules called messenger RNA (mRNA) carry this information out of the nucleus to the molecules that are responsible for synthesizing proteins. The process of producing these mRNA molecules is called transcription.

TRANSCRIPTION

An enzyme called RNA polymerase catalyzes the synthesis of mRNA. DNA provides the template. RNA polymerase first binds to a DNA sequence referred to as a promoter. The promoter positions the enzyme in the correct spot on the correct strand, and guides it in the proper direction to read the code. The process is similar to DNA replication, since RNA polymerase generates a complementary strand and preserves the information in the sequence of the gene. However, the synthesized strand is RNA, and because RNA does not use thymine bases (*see* p. 26), a uracil (U) base replaces thymine in the sequence.

Each gene is located on a specific area of a chromosome. The chromosomes in a cell fall into pairs and each member of a pair carries genes for the same feature.

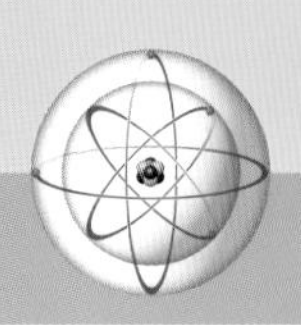

DNA transcription

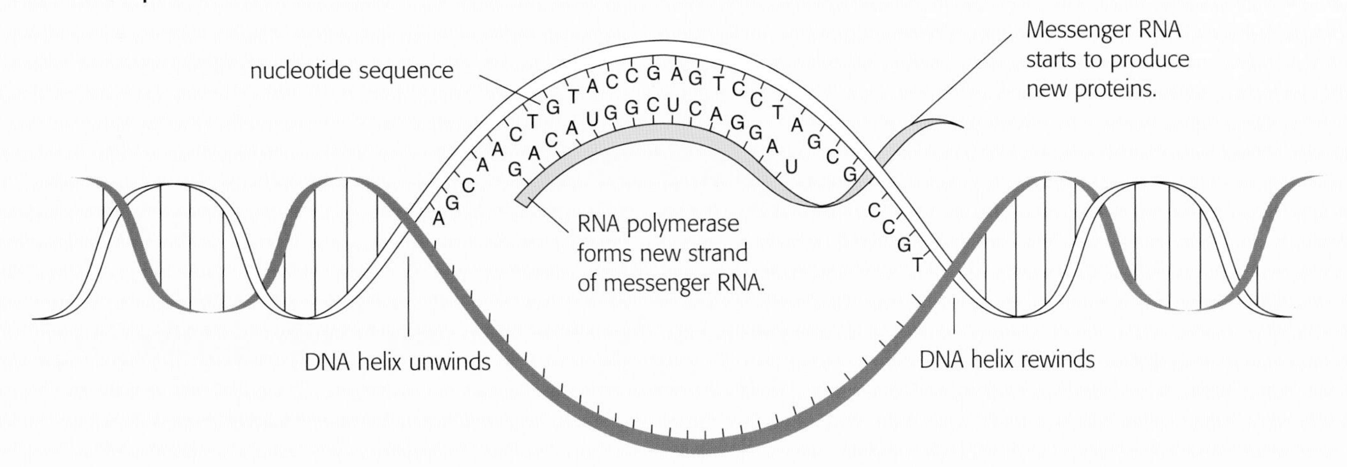

During DNA transcription, a section of the molecule unwinds. As it does, the enzyme RNA polymerase starts to form a strand of messenger RNA that corresponds with the sequence of nucleotide bases on each strand of DNA (only one shown here). After the messenger RNA strand has formed, it moves to the cell's interior. There, the information is translated and used to make new proteins.

Tools and Techniques

Microarrays

Humans have about 25,000–30,000 genes. One way to study a cell is to examine the specific genes that it expresses (transcribes). With thousands of possibilities this can be a daunting task, but tools called microarrays make it easier. A microarray contains a large number of single-stranded DNA genes or gene segments attached to a small glass slide or membrane. When the contents of a cell are poured over the microarray, specific mRNA molecules bind with the corresponding gene on the microarray. Microarrays give scientists a quick method of determining which genes have been expressed, because the location of each gene on the slide is known.

A microarray, sometimes called a biochip, consists of a glass slide holding dots of DNA. Each dot contains a different piece of DNA that will bind to a specific gene in the sample being tested. The microarray is scanned by a laser, which highlights dots that have reacted. The colors indicate which genes are present in the sample.

Chemistry in Action

The problem with genes

Sickle-cell anemia is a disease caused by misshapen hemoglobin, a protein in the blood that carries oxygen. As a result, blood cells are not round but sickle shaped, which slows blood circulation. People who have this disease carry a mutation in both copies of the gene that codes for hemoglobin. The change is small—the misshapen hemoglobin protein is different from normal hemoglobin by only one amino acid. People who have one correct copy of the gene are much less affected, and even show some benefits, since they have a higher resistance to a dangerous disease called malaria.

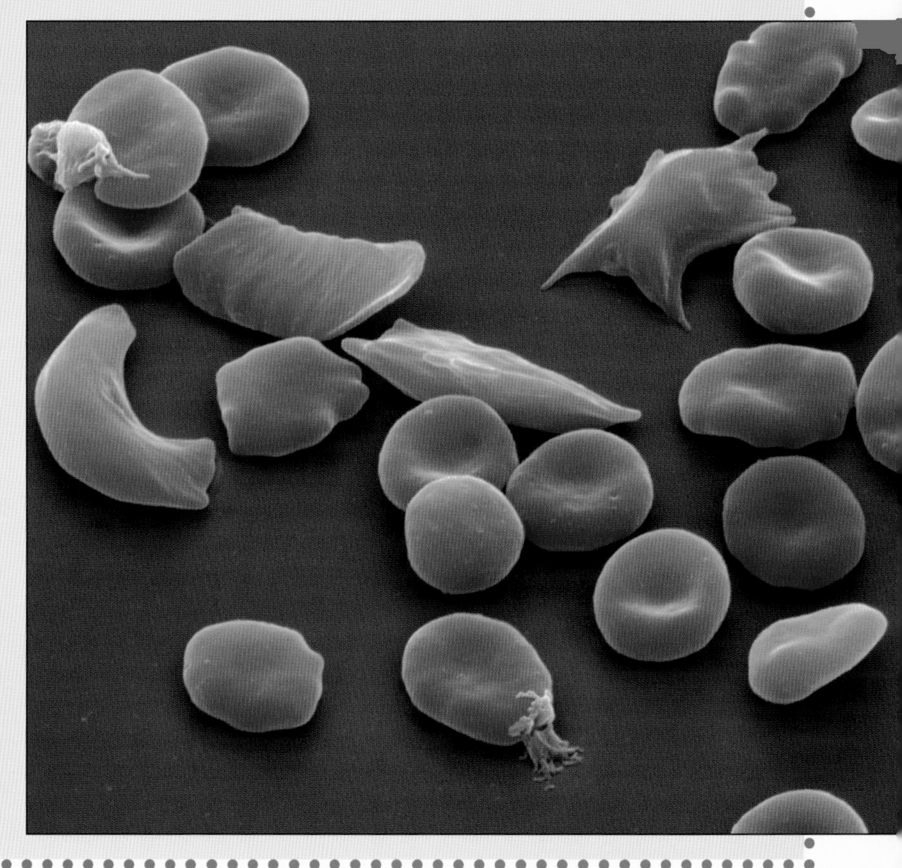

▶ *These blood cells come from someone who has sickle-cell anemia. Normal blood cells are round and red. The sickle cells are pink and irregularly shaped.*

Thus, cells have two forms of each gene, although the two genes may be slightly different versions, or alleles. One allele came from the mother and one from the father. Sometimes cells transcribe only one of these two, and sometimes both, depending on which gene is involved.

Almost all cells in the body contain a full set of chromosome pairs. The exceptions include red blood cells, which do not have any (none are needed in these cells because they have short lifetimes), and reproductive cells such as sperm and eggs, which have only one chromosome of each pair. The union of a sperm and egg during reproduction provides the full set, one of each pair coming from the father's sperm cell and one from the mother's egg cell.

Although cells of the body have the same genes, there are major differences in the cells composing various organs. Cells have special functions. Muscle cells contract, for example, while brain cells transmit messages to one another. These functions depend on the production of proteins that perform the required tasks. Cells from muscle, brain, liver, skin, and

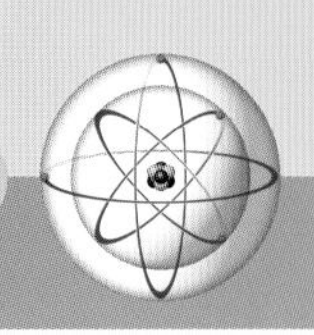

other organs are different not because they have different genes but because they transcribe, or express, a differently ordered set of proteins. The proteins from the set of transcribed genes carry out the cell's special function.

TRANSLATION

The process of synthesizing a protein coded by the mRNA sequence is called translation. Translation occurs in ribosomes, complex cellular structures containing three subunits of ribosomal RNA (rRNA) and dozens of different proteins. Ribosomes translate the mRNA by "reading" each three-base portion of the sequence as a code for the making of a specific amino acid. The amino acids are attached to molecules of transfer RNA (tRNA). Each type of tRNA carries one of the 20 different amino acids. Translation is a complex operation. Enzymes catalyze each step as the ribosome moves along the mRNA, synthesizing the protein according to the coded instruction.

Each three-base segment of the mRNA is called a codon, and it codes for a specific amino acid. This code is called the genetic code, because it is how genes transmit their information. RNA consists of four different bases, so each of the three positions in a codon can have one of four different bases or "letters"—A, C, U, or G. Changing the sequence of bases gives the possibility of 64 different codons or "words" in the genetic code.

Ribosomes accept the mRNA as a correct copy of the gene. If this is not true, the ribosome produces a protein with an incorrect sequence. Such proteins generally do not have the proper shape to perform their function, so mutations in the gene sequence

▼ *Translating mRNA into protein requires a short molecule called transfer RNA (tRNA). This molecule has an anticodon (a complementary sequence of bases to a codon of mRNA) at one end and an amino acid at the other. The ribosomes in the cell link the anticodon on the tRNA to the right codon on the mRNA. The amino acid forms a bond with the amino acid in front of it, then detaches from the tRNA. The tRNA then moves off to pick up a new molecule of its amino acid.*

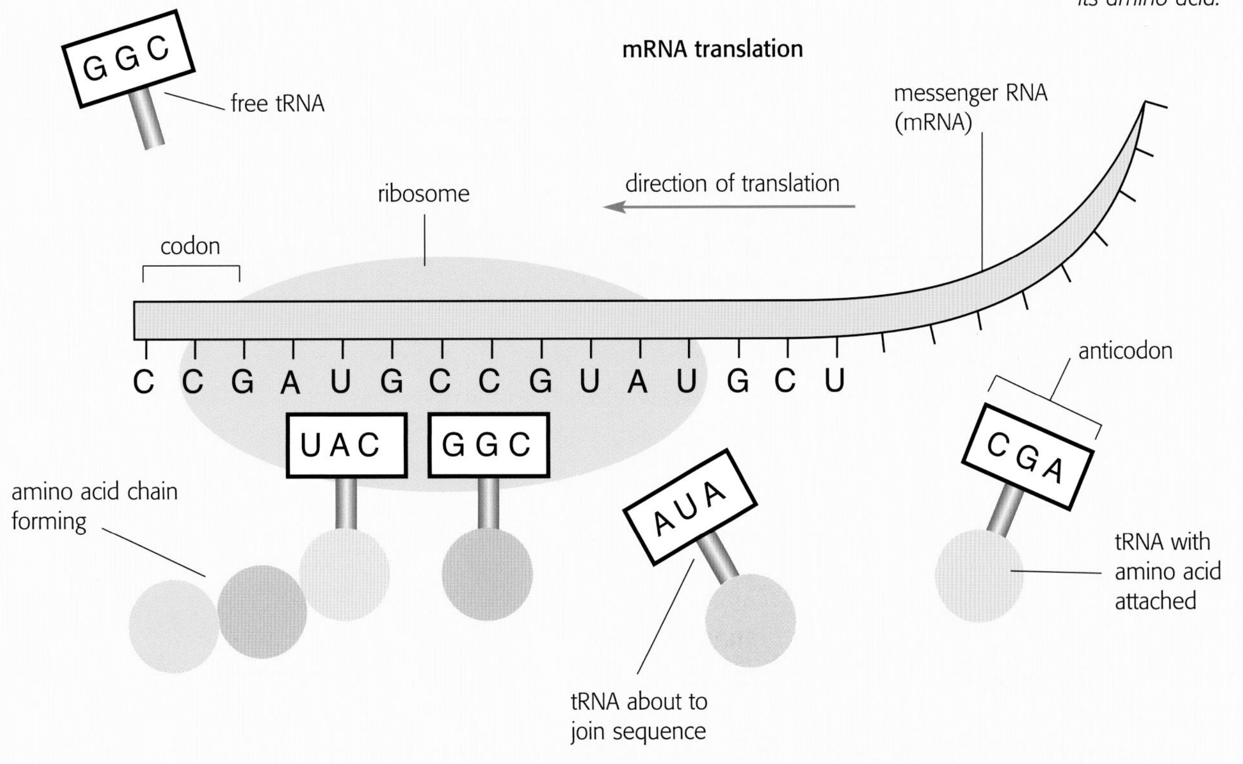

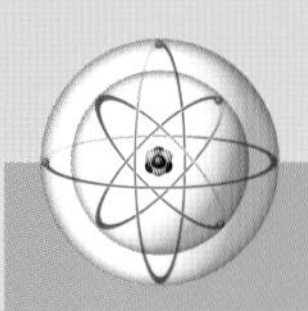

◀ *An electron micrograph of a glycine crystal. Glycine is the simplest amino acid, consisting of one central carbon atom, two hydrogen atoms, one amino group, and one carboxyl group. It is used in the synthesis of the nucleotides adenine and guanine, and is thought to have been the first amino acid to form on Earth.*

produce an incorrect or abnormal protein. On rare occasions the new protein may fulfill a new and beneficial role, but much more often the protein is useless or even harmful.

Each codon represents either an amino acid or a start or stop instruction. When the ribosome encounters the GGG code, it inserts the amino acid glycine into the protein sequence. Enzymes associated with the ribosome catalyze the peptide bond that attaches the glycine to the previous amino acid of the growing chain. The codon AUG codes for the amino acid methionine as well as signaling the beginning of the protein. The codons UAA, UGA, and UAG signal the ribosome to stop translating.

Molecules of tRNA identify the amino acid they carry by having a three-base sequence in their nucleotides that is complementary to the amino acid's codon. This sequence is called an anticodon. For example, the codon AAC codes for asparagine, and a tRNA molecule with the anticodon UUG carries an asparagine amino acid. The anticodon binds to the mRNA codon coding for the proper amino acid, positioning the tRNA so the ribosome can snatch its amino acid and add it to the chain. The tRNA then falls off and picks up another molecule of its amino acid, ready to participate in another reaction.

DNA RECOMBINATION

Cells do all they can to maintain their DNA sequences. However, in some situations there is a benefit to shuffling the genes of the chromosomes, which is called DNA recombination.

A Closer LOOK

Shuffling the gene pack

Shuffling genes is nature's way of ensuring the survival of species. Both parents have two sets of chromosomes but provide only one set of chromosomes to the offspring. Which set the offspring receives is entirely random. The genes on the chromosomes are mixed up again when the DNA is copied. Some are taken from one parent and some from the other. In this way, new adaptations occur that may give the organism an advantage over competitors.

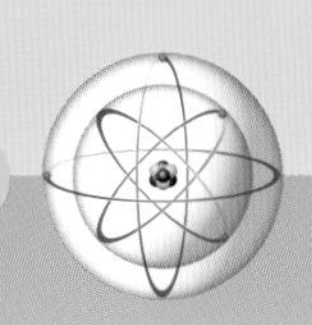

During reproduction, the offspring receives one chromosome of each pair from each parent. When the chromosomes are rejoined in the offspring, the DNA is recombined in such a way that it takes some sections from one parent and some sections from the other parent. This is nature's way of "experimenting," with the goal of producing the best possible combination of genetic features for the survival of the offspring.

DNA recombination requires the breaking of DNA strands. A large segment or just a small portion of one chromosome crosses over to the other of the pair, and then the strands join up with their new partner. The chromosome pair swaps parts, both large and small.

Mixing of genes in recombination occurs naturally, but scientists have learned how to manage a similar procedure. Recombinant DNA technology alters or manipulates the genes of an organism, sometimes producing a new kind of plant or animal.

TRANSGENIC ORGANISMS

Scientists have learned enough about biochemical reactions and how they occur in organisms to perform many of these reactions in the laboratory. We can also use this knowledge to create unique situations and even unique organisms by moving genes and DNA from one species into another.

Organisms having DNA sequences that normally do not appear in their species are called transgenic organisms. The word transgenic means "moving"

Key Terms

- **Allele:** An alternative version of a specified gene.
- **Chromosome:** A long, coiled molecule of DNA that carries genetic information.
- **Codon:** A sequence of three nucleotide bases that codes for an amino acid.
- **Peptide bond:** The bond that forms when two amino acids join.
- **Reproduction:** The biological process by which a new generation of plants or animals is created.

▼ *If you put this pig under an ultraviolet light it will glow in the dark. This happens because scientists have inserted a gene from a fluorescent jellyfish into the pig's DNA, which is why the pig's nose looks slightly green.*

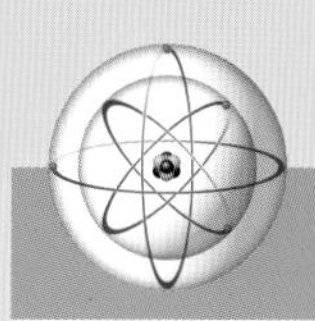

or "transferring" genes from one organism to another.

Examples of transgenic organisms include aquarium fish that glow. Inserted into the DNA of these fish is a gene from another animal, such as a jellyfish, that is capable of bioluminescence. Bioluminescence refers to the emission of light by a living organism, such as a jellyfish. The aquarium fish do not normally have the gene to do this, but the transgenic fish do because the gene has been inserted into one of their chromosomes.

The biochemistry of handling and transferring DNA involves enzymes that cut DNA strands and paste them together again. Enzymes that cut DNA are called restriction enzymes. These enzymes cut DNA at specific sequences and were first discovered in bacteria, where various types of these enzymes seemed to be restricted to certain strains of bacteria (which explains the enzyme name). Scientists and technicians use restriction enzymes to cut DNA at specific places so that the desired sequences can be removed. Enzymes called DNA ligases catalyze the reconnection of DNA strands and are useful when pasting genes or other sequences into a stretch of DNA.

A tool called the polymerase chain reaction (PCR) has also been helpful in many applications of DNA technology. In 1985 Kary Mullis, working for a company called Cetus, invented the polymerase chain reaction (PCR) to multiply small

Profile

Kary Mullis

Kary Mullis was born in 1944 in North Carolina. He earned a Ph.D. in biochemistry in 1972 from the University of California, Berkeley. His invention, the polymerase chain reaction, is used in laboratories all over the world. Mullis received the Nobel Prize in chemistry in 1993, less than a decade after his achievement. Nobel Prizes are not commonly awarded so soon except for accomplishments of great importance.

Kary Mullis, one of the few scientists to receive a Nobel Prize less than 10 years after making his breakthrough discovery.

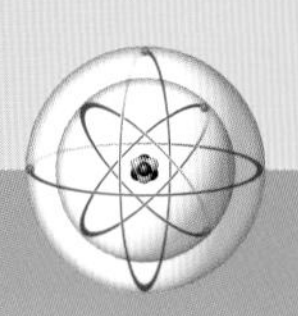

Tools and Techniques

Polymerase chain reaction

Forensic scientists working at a crime scene used to find it difficult to obtain enough biological material on which to carry out DNA tests. However, even the most microscopic samples can be tested using a technique called the polymerase chain reaction (PCR). This method acts like a xerox machine to make multiple copies of a single DNA sequence.

When a cell divides, it uses enzymes called polymerases to copy the DNA in its chromosomes. PCR uses the same method. The first step is to heat the DNA sample so the two strands of DNA unzip. As they separate, a polymerase taken from a high-temperature bacteria makes a copy using each strand as a template. However, the polymerase cannot begin to copy the strand unless a short sequence of nucleotides is already present to start the process. Another enzyme is used as a primer to build this first sequence, but the sample has to be cooled for this to happen. The temperature is then raised again to allow the polymerase to copy the sequence.

This cycle, which takes about two minutes, is repeated up to 30 times. Because each copied sequence can act as a new template, at the end of the process there are 1 billion new pieces of DNA. These pieces can then be used as markers to look for similar fragments in other samples. If the pieces match, then the forensic scientists know they have a likely suspect.

▲ *This cigarette butt is being tested for traces of DNA left on it by the smoker's saliva. From such a small sample as this, scientists can discover the DNA of the person who smoked the cigarette.*

portions of DNA (*see* box above). PCR copies selected DNA strands by cycles of heating and cooling. The heating separates the double helix into single strands so that they can be copied. (Cells likewise have to "unzip" DNA to replicate it.) Lowering the temperature allows enzymes to bind to the strand and replicate it. The enzymes involved in DNA replication are called DNA polymerases, but those used in the PCR process are taken from a bacteria that can withstand high temperatures. Unlike most proteins, these bacterial DNA polymerases are not destroyed by heat and can endure the heating cycles. PCR generates enough copies of the desired DNA sequence to allow scientists to modify or experiment with specific sections of genetic information.

The potential benefits of genetic modification include adjusting the genes

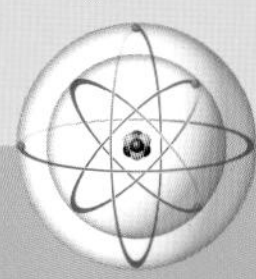

of crops such as corn or tomatoes to grow faster, stay fresher longer, or resist insect attack. Genetically modified food could increase the world's food supply and feed an ever growing population. An increasing number of crops have genetically modified seeds that farmers can purchase from companies. However, some people worry that "engineering" genetic information in this way could have unpredictable consequences. Their concern is that genetically modified organisms could disrupt the balance of nature, causing damage to the environment and a possible loss of genetic variation. Since variation is crucial to evolution and the adaptability of organisms, some people fear that vulnerable but important species may be lost. This debate is not yet resolved.

GENE THERAPY

Biochemical techniques that modify genetic information include the possibility of correcting genetic defects. Gene therapy is a procedure in which a correct copy of a gene is used to replace a mutated or dysfunctional one. Although gene therapy is not yet a common treatment option for genetic diseases, researchers are working on developing and refining this technique.

Researchers have been modifying the genes of laboratory animals, such as mice, for several decades. Scientists delete or "knock out" a gene from the reproductive cells of mice and their

◀ *Genetically modified crops, such as these tomatoes, are just one example of what can be achieved by manipulating an organism's DNA.*

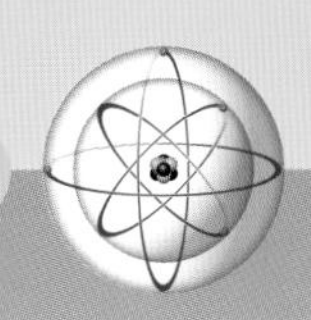

Chemistry in Action

Gene therapy

The retina, which is the lining of the back of the eye, contains cells that turn light into electrical impulses, a process crucial in vision. In diseases such as retinitis pigmentosa, a mutation in one of a number of different genes leads to blindness. Physicians and scientists are working on gene therapy techniques to treat this and other disorders. In the last few years progress has been made in overcoming this problem. Researchers have been able to restore partial eyesight to dogs and mice suffering from a type of genetic disorder related to the human disease Leber congenital amaurosis, which also causes blindness.

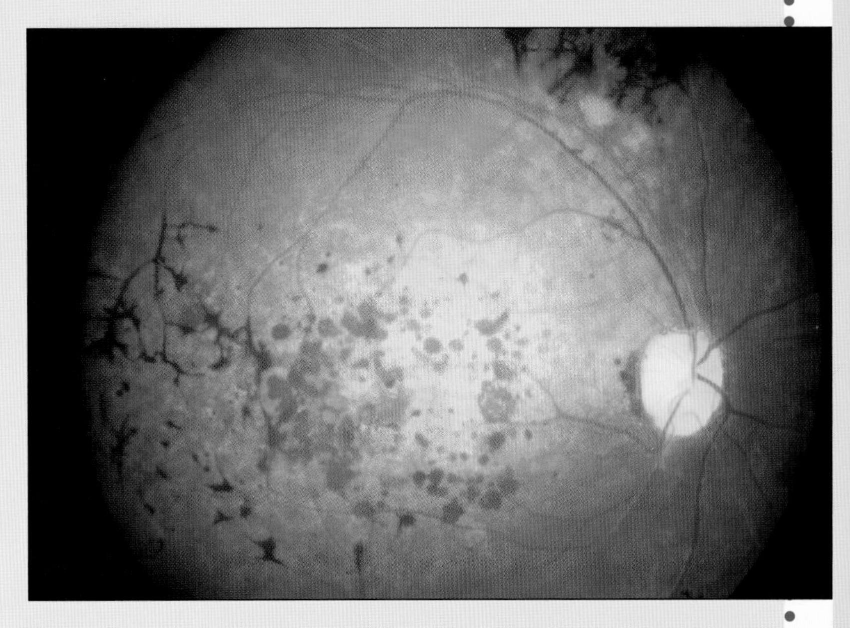

▲ *A retina showing the dark areas where sight has been lost through retinitis pigmentosa.*

offspring using DNA recombination. However, the new inserted sequence, which replaces the normal gene, is not functional. Using this method, scientists can determine the function of a gene by comparing the characters of organisms that are genetically identical except that the "knock out" mouse is missing the gene. Gene therapy is more difficult since medical treatments cannot be as intrusive, and usually involve only a specific type and number of cells of the patient.

◀ *Knock out mice are used in laboratories to test the functions of different genes.*

See ALSO ...
- *Metabolic Pathways, Vol. 9: pp. 28–43.*

More Information

BOOKS

Atkins, P. W. *The Periodic Kingdom: A Journey into the Land of Chemical Elements.* New York, NY: Basic Books, 1997.

Bendick, J., and Wiker, B. *The Mystery of the Periodic Table (Living History Library).* Bathgate, ND: Bethlehem Books, 2003.

Berg, J., Stryer, L., and Tymoczko, J. *Biochemistry.* New York, NY: W. H. Freeman, 2002.

Brown, T., Burdge, J., Bursten, B., and LeMay, E. *Chemistry: The Central Science.* 10th ed. Englewood Cliffs, NJ: Prentice Hall, 2005.

Cobb, C., and Fetterolf, M. L. *The Joy of Chemistry: The Amazing Science of Familiar Things.* Amherst, NY: Prometheus Books, 2005.

Cox, M., and Nelson, D. *Lehninger's Principles of Biochemistry.* 4th ed. New York, NY: W. H. Freeman, 2004.

Davis, M. *Modern Chemistry.* New York, NY: Henry Holt, 2000.

Herr, N., and Cunningham, J. *Hands-on Chemistry Activities with Real Life Applications.* Hoboken, NJ: Jossey-Bass, 2002.

Houck, Clifford C., and Post, Richard. *Chemistry: Concepts and Problems.* Hoboken, NJ: Wiley, 1996.

Karukstis, K. K., and Van Hecke, G. R. *Chemistry Connections: The Chemical Basis of Everyday Phenomena.* Burlington, MA: Academic Press, 2003.

LeMay, E. *Chemistry: Connections to Our Changing World.* New York, NY: Prentice Hall (Pearson Education), 2000.

Oxlade, C. *Elements and Compounds.* Chicago, IL: Heinemann, 2002.

Poynter, M. *Marie Curie: Discoverer of Radium* ***(Great Minds of Science).*** Berkeley Heights, NJ: Enslow Publishers, 2007.

Saunders, N. *Fluorine and the Halogens.* Chicago, IL: Heinemann Library, 2005.

Shevick, E., and Wheeler, R. *Great Scientists in Action: Early Life, Discoveries, and Experiments.* Carthage, IL: Teaching and Learning Company, 2004.

Stwertka, A. *A Guide to the Elements.* New York, NY: Oxford University Press, 2002.

Tiner, J. H. *Exploring the World of Chemistry: From Ancient Metals to High-Speed Computers.* Green Forest, AZ: Master Books, 2000.

Trombley, L., and Williams, F. *Mastering the Periodic Table: 50 Activities on the Elements.* Portland, ME: Walch, 2002.

Walker, P., and Wood, E. *Crime Scene Investigations: Real-life Science Labs for Grades 6–12.* Hoboken, NJ: Jossey-Bass, 2002.

Wertheim, J. *Illustrated Dictionary of Chemistry* (Usborne Illustrated Dictionaries). Tulsa, OK: Usborne Publishing, 2000.

Wilbraham, A., et al. *Chemistry.* New York, NY: Prentice Hall (Pearson Education), 2000.

Woodford, C., and Clowes, M. *Routes of Science: Atoms and Molecules.* San Diego, CA: Blackbirch Press, 2004.

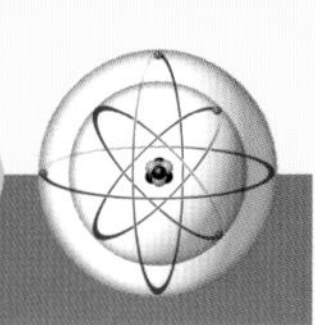

WEB SITES

The Art and Science of Bubbles
www.sdahq.org/sdakids/bubbles
Information and activities about bubbles.

Chemical Achievers
www.chemheritage.org/classroom/chemach/index.html
Biographical details about leading chemists and their discoveries.

The Chemistry of Batteries
www.science.uwaterloo.ca/~cchieh/cact/c123/battery.html
Explanation of how batteries work.

The Chemistry of Chilli Peppers
www.chemsoc.org/exemplarchem/entries/mbellringer
Fun site giving information on the chemistry of chilli peppers.

The Chemistry of Fireworks
library.thinkquest.org/15384/chem/chem.htm
Information on the chemical reactions that occur when a firework explodes.

The Chemistry of Water
www.biology.arizona.edu/biochemistry/tutorials/chemistry/page3.html
Chemistry of water and other aspects of biochemistry.

Chemistry: The Periodic Table Online
www.webelements.com
Detailed information about elements.

Chemistry Tutor
library.thinkquest.org/2923
A series of Web pages that help with chemistry assignments.

Chem4Kids
www.chem4Kids.com
Includes sections on matter, atoms, elements, and biochemistry.

Chemtutor Elements
www.chemtutor.com/elem.htm
Information on a selection of the elements.

Eric Weisstein's World of Chemistry
scienceworld.wolfram.com/chemistry
Chemistry information divided into eight broad topics, from chemical reactions to quantum chemistry.

General Chemistry Help
chemed.chem.purdue.edu/genchem
General information on chemistry plus movie clips of key concepts.

Molecular Models
chemlabs.uoregon.edu/GeneralResources/models/models.html
A site that explains the use of molecular models.

New Scientist
www.newscientist.com/home.ns
Online science magazine providing general news on scientific developments.

Periodic Tables
www.chemistrycoach.com/periodic_tables.htm#Periodic%20Tables
A list of links to sites that have information on the periodic table.

The Physical Properties of Minerals
mineral.galleries.com/minerals/physical.htm
Methods for identifying minerals.

Understanding Our Planet Through Chemistry
minerals.cr.usgs.gov/gips/aii-home.htm
Site that shows how chemists and geologists use analytical chemistry to study Earth.

Scientific American
www.sciam.com
Latest news on developments in science and technology.

Snowflakes and Snow Crystals
www.its.caltech.edu/~atomic/snowcrystals
A guide to snowflakes, snow crystals, and other ice phenomena.

Virtual Laboratory: Ideal Gas Laws
zebu.uoregon.edu/nsf/piston.html
University of Oregon site showing simulation of ideal gas laws.

What Is Salt?
www.saltinstitute.org/15.html
Information on common salt.

Periodic Table

The periodic table organizes all the chemical elements into a simple chart according to the physical and chemical properties of their atoms. The elements are arranged by atomic number from 1 to 116. The atomic number is based on the number of protons in the nucleus of the atom. The atomic mass is the combined mass of protons and neutrons in the nucleus. Each element has a chemical symbol that is an abbreviation of its name. In some cases, such as potassium,

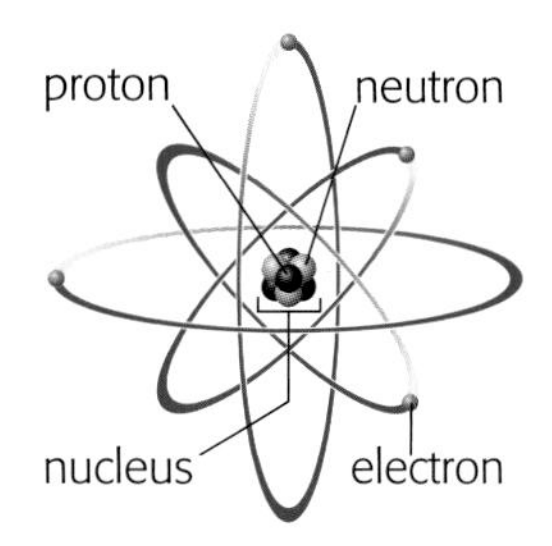

Atomic structure

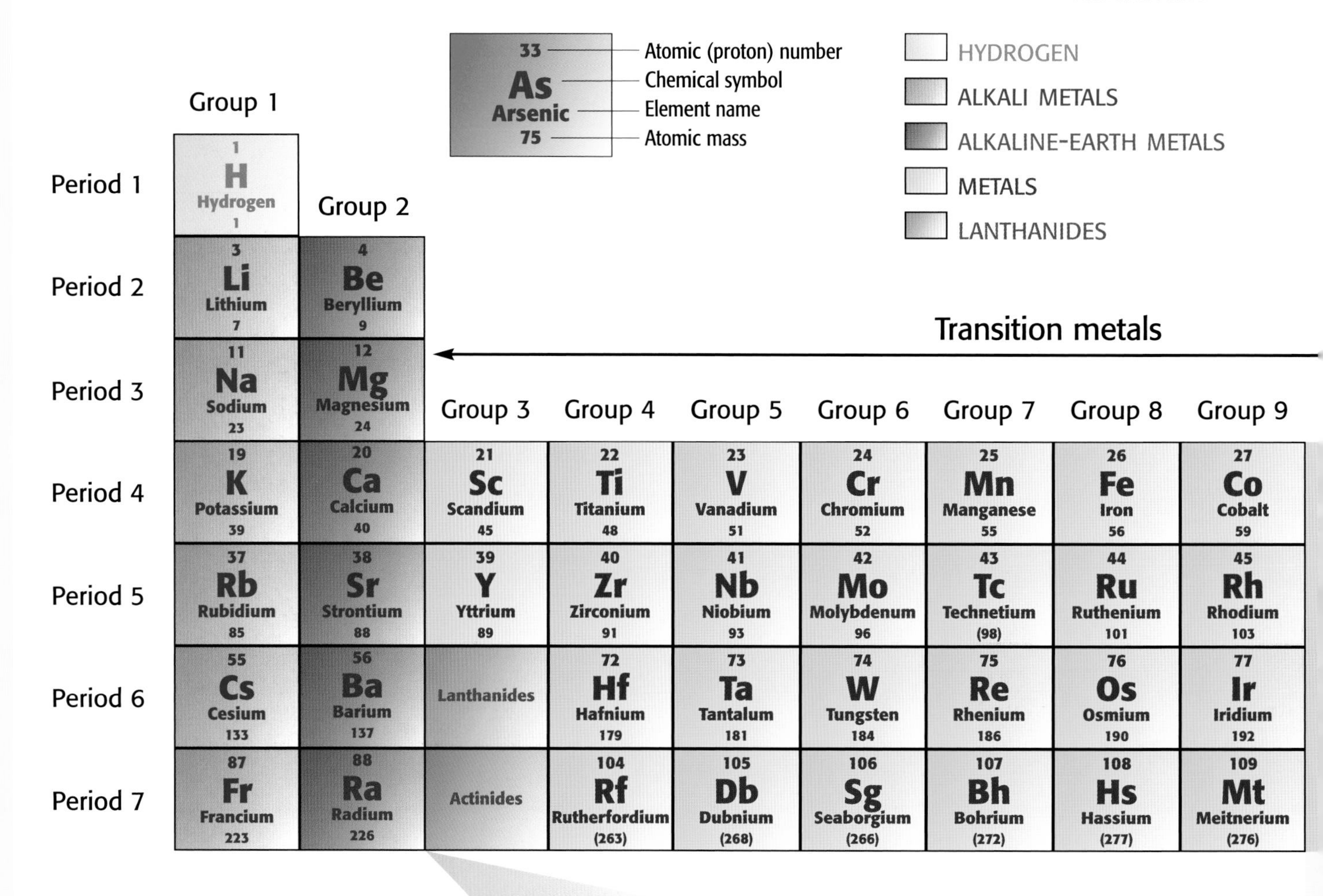

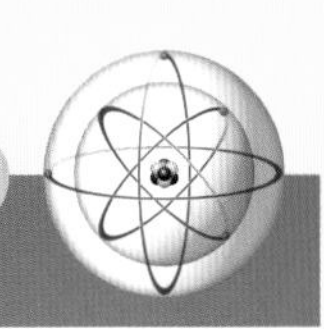

the symbol is an abbreviation of its Latin name ("K" stands for *kalium*). The name by which the element is commonly known is given in full underneath the symbol. The last item in the element box is the atomic mass. This is the average mass of an atom of the element.

Scientists have arranged the elements into vertical columns called groups and horizontal rows called periods. Elements in any one group all have the same number of electrons in their outer shell and have similar chemical properties. Periods represent the increasing number of electrons it takes to fill the inner and outer shells and become stable. When all the spaces have been filled (Group 18 atoms have all their shells filled) the next period begins. Further explanation of the periodic table is given in Volume 5.

ACTINIDES
NOBLE GASES
NONMETALS
METALLOIDS

Group 10	Group 11	Group 12	Group 13	Group 14	Group 15	Group 16	Group 17	Group 18
								2 **He** Helium 4
			5 **B** Boron 11	6 **C** Carbon 12	7 **N** Nitrogen 14	8 **O** Oxygen 16	9 **F** Fluorine 19	10 **Ne** Neon 20
			13 **Al** Aluminum 27	14 **Si** Silicon 28	15 **P** Phosphorus 31	16 **S** Sulfur 32	17 **Cl** Chlorine 35	18 **Ar** Argon 40
28 **Ni** Nickel 59	29 **Cu** Copper 64	30 **Zn** Zinc 65	31 **Ga** Gallium 70	32 **Ge** Germanium 73	33 **As** Arsenic 75	34 **Se** Selenium 79	35 **Br** Bromine 80	36 **Kr** Krypton 84
46 **Pd** Palladium 106	47 **Ag** Silver 108	48 **Cd** Cadmium 112	49 **In** Indium 115	50 **Sn** Tin 119	51 **Sb** Antimony 122	52 **Te** Tellurium 128	53 **I** Iodine 127	54 **Xe** Xenon 131
78 **Pt** Platinum 195	79 **Au** Gold 197	80 **Hg** Mercury 201	81 **Tl** Thallium 204	82 **Pb** Lead 207	83 **Bi** Bismuth 209	84 **Po** Polonium (209)	85 **At** Astatine (210)	84 **Rn** Radon (222)
110 **Ds** Darmstadtium (281)	111 **Rg** Roentgenium (280)	112 **Uub** Ununbium (285)	113 **Uut** Ununtrium (284)	114 **Uuq** Ununquadium (289)	115 **Uup** Ununpentium (288)	116 **Uuh** Ununhexium (292)		

artificial elements

62 **Sm** Samarium 150	63 **Eu** Europium 152	64 **Gd** Gadolinium 157	65 **Tb** Terbium 159	66 **Dy** Dysprosium 163	67 **Ho** Holmium 165	68 **Er** Erbium 167	69 **Tm** Thulium 169	70 **Yb** Ytterbium 173	71 **Lu** Lutetium 175
94 **Pu** Plutonium (244)	95 **Am** Americium (243)	96 **Cm** Curium (247)	97 **Bk** Berkelium (247)	98 **Cf** Californium (251)	99 **Es** Einsteinium (252)	100 **Fm** Fermium (257)	101 **Md** Mendelevium (258)	102 **No** Nobelium (259)	103 **Lr** Lawrencium (260)

Glossary

acid Substance that dissolves in water to form hydrogen ions (H^+). Acids are neutralized by alkalis and have a pH below 7.

active site The area of an enzyme where the reactants attach and the reaction occurs.

adenine A purine base that is a constituent of both RNA and DNA.

alkali Substance that dissolves in water to form hydroxide ions (OH^-). Alkalis have a pH greater than 7 and will react with acids to form salts.

allele An alternative form of a specified gene.

amino acid A component of a protein made up of a carboxyl group (–COOH) and an amino group ($-NH_2$), both attached to the same carbon atom.

anesthetic A substance that causes lack of feeling or awareness in the body.

antibiotic A substance that destroys or slows the growth of microorganisms, including harmful bacteria and fungi.

antiseptic A substance that kills or restricts the growth of harmful germs.

atom The smallest independent building block of matter. All substances are made of atoms.

atomic mass number The number of protons and neutrons in an atom's nucleus.

atomic number The number of protons in a nucleus.

ATP (adenosine triphosphate) A substance that carries chemical energy in organisms. It consists of a base, a sugar, and three phosphate groups.

bacteria A very varied group of one-celled microorganisms.

base Any substance that produces hydroxide ions (OH^-) is a base. All alkalis are bases.

base pair A base and its partner in a strand of DNA or RNA. These are always adenine and thymine (or uracil), and guanine and cytosine.

bond A chemical connection between atoms.

carbohydrate One of a group of compounds that includes sugars, starch, and cellulose. Some are essential in turning food to energy. Others are energy stores in plants, and still more build plant and animal cell membranes.

catalyst Substance that speeds up a chemical reaction but is left unchanged at the end of the reaction.

cell membrane Thin sheet that holds in the contents of a cell.

cellulose Polysaccharide made from chains of glucose molecules. Cellulose serves as a structural component of cell walls in plants.

chemical equation Symbols and numbers that show how reactants change into products during a chemical reaction.

chemical formula The letters and numbers that represent a chemical compound, such as "H_2O" for water.

chemical reaction The reaction of two or more chemicals (the reactants) to form new chemicals (the products).

chemical symbol The letters that represent a chemical, such as "Cl" for chlorine or "Na" for sodium.

chiral molecules Isomer molecules that are a mirror image of each other.

cholesterol Molecule from which steroids are derived. Excess of this molecule in the body may cause heart disease.

chromosome A long, coiled molecule of DNA that carries genetic (inherited) information.

codon A sequence of three nucleotide bases that codes for an amino acid.

compound Substance made from more than one element and that has undergone a chemical reaction.

condensation reaction A reaction that produces water, such as when a glycosidic bond is made.

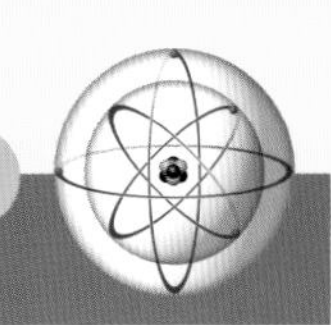

covalent bond Bond in which atoms share one or more electrons.

cytosine Pyramidine base that is a constituent of both RNA and DNA.

deoxyribonucleic acid (DNA) Molecule that contains the genetic code for all cellular organisms.

dipole attraction The attractive force between the electrically charged ends of molecules.

disaccharide Carbohydrate made of two saccharide molecules.

dissolve To form a solution.

double helix The shape of a DNA molecule. It resembles a twisted ladder.

electron A tiny, negatively charged particle that moves around the nucleus of an atom.

enzyme A biological protein that acts as a catalyst.

fatty acid A compound made up of a hydrocarbon chain and a carboxyl group. The shortest chain has just one hydrogen atom, but the longest has nearly 30 carbon atoms.

functional group A section of an organic molecule that gives it certain chemical properties.

galactose Monosaccharide that combines with glucose to make the disaccharide lactose, the sugar found in milk.

glucose Monosaccharide also known as dextrose or grape sugar. An important energy source in animals and plants.

glycogen Polysaccharide made of glucose. Animals store glucose in the form of glycogen.

glycosidic bond A bond that connects one saccharide to another saccharide.

hemoglobin A protein that is used by many animals to transport oxygen in blood.

hormone Molecules that travel in the bloodstream and carry messages from cells in one part of the body to another.

hydrogen bond A weak dipole attraction that always involves a hydrogen atom.

hydrolysis The process by which a molecule splits after reacting with a molecule of water.

hydrophilic Describes a thing that has an attraction to water.

hydrophobic Describes a thing that does not have an attraction to water.

hydroxyl A functional group (–OH) made up of an oxygen and hydrogen atom.

inorganic A compound that is not organic.

insoluble A substance that cannot dissolve.

intermolecular bonds The bonds that hold molecules together. These bonds are weaker than those between atoms in a molecule.

intramolecular bond Strong bond between atoms in a molecule.

ion An atom that has lost or gained one or more electrons.

ionic bond Bond in which one atom gives one or more electrons to another atom.

ionization The formation of ions by adding or removing electrons from atoms.

isomer Compounds with the same chemical formula but different arrangements of their atoms.

isotope Atoms of a given element always have the same number of protons but can have different numbers of neutrons. These different versions of the same element are called isotopes.

keratin A type of protein that forms structures such as fingernails and horns.

lactase Enzyme that breaks down lactose (the disaccharide in milk) into glucose and galactose.

lipids Organic compounds that are insoluble in water. Fats and oils are lipids.

metabolism The chemical reactions that occur within living organisms and that are necessary for its function.

mitochondria Cell structures that produce ATP.

monosaccharide Carbohydrate made of one saccharide molecule.

neutron One of the particles that make up the nucleus of an atom. Neutrons do not have an electric charge.

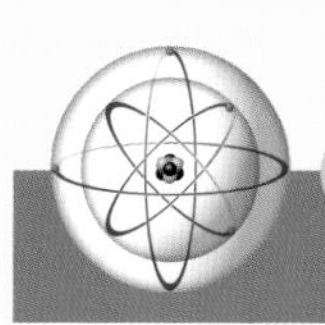

nitrogen fixation Processes that combine atmospheric nitrogen with other elements in a form that can be absorbed by plants.

nonpolar solvent Solvents in which the molecules are made of atoms that share electrons equally, so there is no excess charge.

nucleus The central part of an atom. The nucleus contains protons and neutrons. The exception is hydrogen, which contains only one proton.

nucleic acid A chain of nucleotides.

nucleotide A molecule made up of a nitrogen-containing base, a sugar, and a phosphate group.

nutrient A substance that provides nourishment to a cell and helps it grow or repair itself.

organic A compound that is made of carbon and hydrogen.

peptide Two bonded amino acids.

peptide bond Bond that forms when two amino acids join.

phospholipid A lipid containing a phosphate group ($-PO_4$). Phospholipids form the membranes of cells.

photon A particle that carries a quantity of energy, such as in the form of light.

photosynthesis A chemical reaction in which plants use energy from the Sun to change carbon dioxide and water into food.

polymerase An enzyme that creates a polymer (long chain) of nucleotides from a template of DNA or RNA.

polysaccharide Carbohydrate made of many saccharide molecules.

protein A substance consisting of large molecules made up of chains of chemical units called amino acids.

proton A positively charged particle found in an atom's nucleus.

purine A component of a nucleotide. Purines have a five-atom ring structure connected to a six-atom ring. Purines include adenine and guanine.

pyrimidine A component of a nucleotide. Pyramidines have a six-atom ring. Pyramidines include thymine, uracil, and cytosine.

relative atomic mass A measure of the mass of an atom compared with the mass of another atom. The values used are the same as those for atomic mass.

ribonucleic acid (RNA) Molecule that carries genetic information. It controls how a cell functions and reproduces.

saccharide A sugar molecule.

solute A substance that dissolves in a solvent.

solution A mixture of two or more elements or compounds in a single phase, or state (solid, liquid, or gas).

solvent The liquid that dissolves a solute.

starch A polysaccharide made of glucose molecules. Starch is the primary molecule that plants use to store carbohydrates for future use.

steroids Lipids that are often hormones. Steroids can pass easily through cell membranes.

subatomic particles Particles that are smaller than an atom.

substrate The particular molecule on which an enzyme acts.

sulfa drugs Organic compounds that are used as antibacterial agents by preventing bacteria from reproducing.

thymine Pyrimidine that is a component of DNA.

triglyceride A major component of fats, triglycerides provide a concentrated food energy store in organisms.

uracil Pyrimidine that is a component in RNA.

viruses Extremely small organisms, which consist of an RNA or DNA core surrounded by a protein covering.

vitamin A substance that is essential for good health.

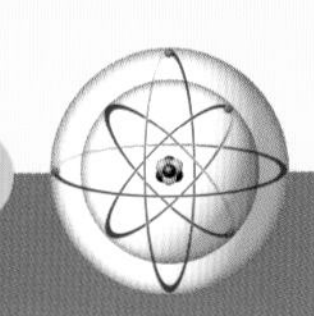

Index

Numbers in **bold** type refer to volume numbers.

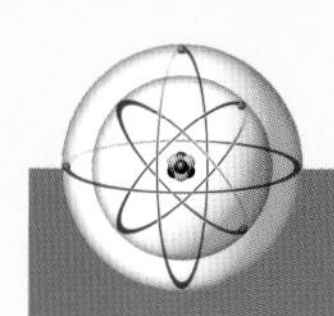

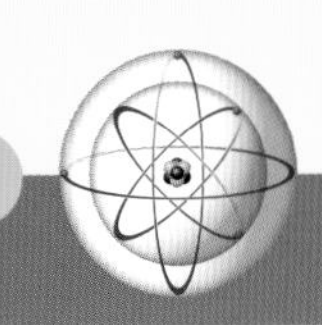

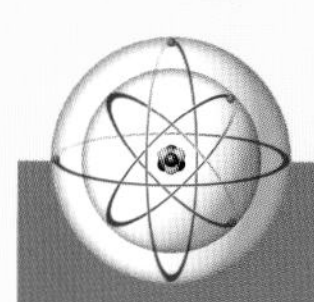

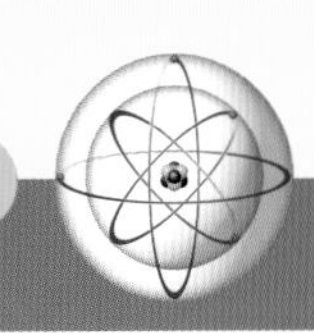

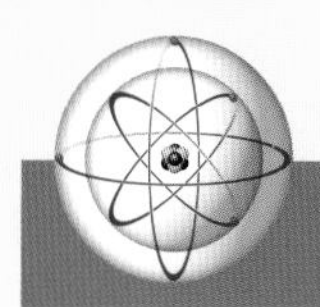

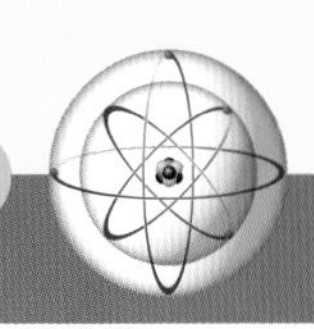

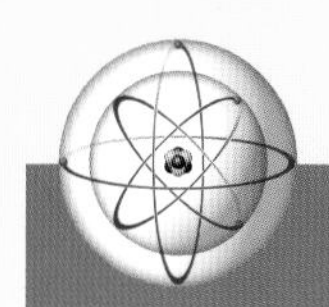